Laser-generated Ultrasound Technology

IEEE Press
445 Hoes Lane
Piscataway, NJ 08854

Laser-generated Ultrasound Technology

Benpeng Zhu
Huazhong University of Science and Technology,
School of Integrated Circuits, Wuhan, China

Di Ke
Huazhong University of Science and Technology,
School of Integrated Circuits, Wuhan, China

Tao Zhang
Huazhong University of Science and Technology,
School of Integrated Circuits, Wuhan, China

IEEE Press
WILEY

Published by John Wiley & Sons, Inc., Hoboken, New Jersey.

Library of Congress Cataloging-in-Publication Data Applied for:

Hardback ISBN: 9781394365098

Cover Design: Wiley
Cover Image: © Mihail/stock.adobe.com

Set in 9.5/12.5pt STIXTwoText by Straive, Chennai, India

Contents

Author Biography

Benpeng Zhu is a Professor at the School of Integrated Circuits, Huazhong University of Science and Technology. In 2024, he was appointed Distinguished Professor under the Changjiang Scholars Program by the Ministry of Education. He currently serves as Vice Chairman of the Medical Ultrasound Engineering Committee of the Acoustical Society of China. He has made significant contributions to the interdisciplinary field of acoustic physics and biomedical engineering, particularly in the development of advanced biomedical ultrasound technologies and intelligent healthcare systems.

Di Ke is a Ph.D. student at the School of Integrated Circuits, Huazhong University of Science and Technology.

Tao Zhang is an Assistant Professor at the School of Integrated Circuits, Huazhong University of Science and Technology.

1

From Photoacoustic Effect to Laser-generated Ultrasound Technology

Ultrasound is a mechanical wave characterized by a frequency beyond the upper threshold of human hearing (20 kHz) and wavelengths typically falling within the millimeter to micrometer range. Methods for generating ultrasonic waves primarily include the piezoelectric method and the magnetostrictive method. The piezoelectric method represents the most widely adopted mainstream technology, operating on the principle of the inverse piezoelectric effect exhibited by piezoelectric crystals such as piezoelectric ceramics and quartz [1]. When subjected to a high-frequency alternating electric field, the crystal lattice of the piezoelectric material undergoes periodic deformation, thereby inducing mechanical vibrations. These vibrations are subsequently transmitted into the propagation medium via a coupling layer to form ultrasonic waves. Conversely, the magnetostrictive method relies on the periodic volumetric changes of ferromagnetic materials—including nickel alloys and ferrites—under an applied alternating magnetic field to excite mechanical vibrations [2].

Despite the established advantages of both piezoelectric and magnetostrictive approaches in ultrasound generation, devices based on electric or magnetic actuation generally exhibit limitations such as narrow operating frequency bands and susceptibility to electromagnetic interference. In recent years, laser-generated ultrasound (LGUS) devices, which operate on the photoacoustic effect, have gained increasing attention due to their unique advantages [3, 4]. As a novel approach for ultrasound generation, LGUS is governed by the fundamental principle of mutual conversion between light energy and acoustic energy. The key operational sequence comprises the absorption of light energy, followed by localized thermoelastic expansion, which ultimately leads to the emission of ultrasonic waves [5].

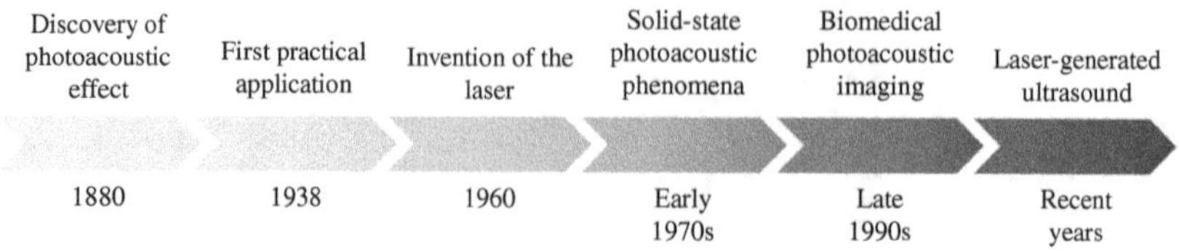

Figure 1.1 The development history from photoacoustic effect to LGUS.

LGUS devices are relatively straightforward to fabricate and possess notable characteristics including broad bandwidth, elimination of the need for direct electrical connections, and inherent immunity to electromagnetic interference [6]. These attributes confer immense promise upon LGUS within the biomedical field, while simultaneously underscoring its versatility for a broad spectrum of applications across diverse scientific and industrial sectors, particularly in scenarios demanding noninvasiveness, high-resolution imaging, or operation in electromagnetically sensitive environments [7].

To fully understand the technological trajectory of LGUS, it is necessary to look back at the origins of its fundamental mechanism. The photoacoustic effect, a distinct physical phenomenon that bridges optical and acoustic principles, has followed an evolutionary path spanning over a century. From its discovery in the late 19th century, through cycles of resurgence and theoretical development in the 20th century, to its active adoption in areas such as biomedicine in the 21st century, its progression illustrates both the persistence and complexity of scientific inquiry, as well as the dynamic potential of cross-disciplinary collaboration [8]. This chapter will track the chronological development of the photoacoustic effect, offering a systematic overview of its full historical trajectory (Figure 1.1) and examining critical innovations, technical evolutions, and expanding applications across distinct periods.

1.1 Discovery of the Photoacoustic Effect

The latter half of the 19th century was a golden era of rapid advancement in modern physics. Progress in electromagnetic theory, innovations in optical technology, and deepening studies in acoustics together established a robust scientific foundation for the eventual discovery of the photoacoustic effect. During this period, scientific exploration often grew from a dynamic interplay between serendipity and deliberate inquiry. It was within this context that the work of Alexander Graham Bell provided the direct catalyst for the emergence of the photoacoustic phenomenon. Motivated by the discovery in 1873 of selenium's photosensitivity—its electrical resistance varies with light intensity—he conceived the idea of using selenium to transform optical signals into electrical ones, thereby achieving wireless vocal communication [9]. After his groundbreaking invention of the telephone in 1876, Bell turned his focus to the possibility of transmitting speech using light [10, 11].

After two years of experimental effort, Bell and his assistant Charles Sumner Tainter succeeded in June 1880 in transmitting wireless voice communication over a distance of approximately 213 meters. Their apparatus, named the "photophone," consisted of a transmitter and a receiver [11]. In operation, sound waves induced vibrations in a glass mirror, thereby modulating the intensity of the reflected light. A selenium cell then converted this modulated optical signal back into an electrical signal, which was ultimately transformed into audible sound via a telephone receiver circuit. Widely recognized as the world's first practical wireless communication device, this invention also laid essential technical groundwork for the subsequent advancement of optical communications [12–14].

Of particular note is a serendipitous discovery made during the research on the photophone: when a solid material was irradiated with a rapidly interrupted beam of light, sound waves were generated at the same frequency as the light modulation [11]. Bell described this phenomenon in a letter to his father, and this observation represents the earliest recognized form of what is now known as the photoacoustic effect.

1.2 The Emergence of Photoacoustic Technology

Following Bell's discovery of the photoacoustic phenomenon, a considerable number of scientists dedicated themselves to exploring its fundamental mechanisms. In 1881, Lord Rayleigh introduced the "solid vibration theory," proposing that intermittent illumination induced non-uniform heating in diaphragms or plate-like structures within objects, resulting in vibrations that generated sound. This theory garnered endorsement from Bell [13, 14]. Concurrently, based on their experimental investigations, Mercadier and Preece advanced the "gas expansion theory," which attributed the origin of sound to the periodic expansion and contraction of air in contact with heated objects [15–17]. Subsequent contributions by Röntgen and Tyndall independently corroborated and further developed the gas expansion theory [18, 19]. Experimental evidence confirmed that the phenomenon also manifested in gases and vapors: Röntgen examined the photoacoustic effect in coal gas and ammonia, while Tyndall systematically studied various gases and vapors, even proposing that the effect held potential for detecting trace amounts of combustible gases in mining environments [19]. Bell further showcased the "spectrophone," an instrument capable of reflecting a sample's absorption spectrum, thereby establishing new pathways for spectral analysis [14]. However, constrained by limitations such as the lack of detection methods capable of precise quantification, research into the photoacoustic effect entered a phase of stagnation. It was not until the period spanning the 1930s to the 1960s that the effect underwent its first substantial revival, notably within the domain of gas analysis.

In the 1930s, Veingerov built upon the earlier work of Röntgen and Tyndall in the study of gas-phase photoacoustic effects [20], particularly emphasizing Tyndall's concept of "utilizing this effect for trace gas detection." In 1938, he introduced a photoacoustic gas analysis method: employing a capacitive microphone as the detector and a Nernst glower as a high-intensity blackbody infrared source, then he successfully measured the concentration of CO_2 in a nitrogen (N_2) matrix [21]. This achievement marked the formal transition of the photoacoustic effect from theoretical exploration to practical application. Over the subsequent years, related technologies progressed rapidly and eventually reached the stage of commercialization [22]. As the range of applications broadened, the theoretical framework of the photoacoustic effect continued to be refined. Beyond its use in gas concentration analysis, it was also applied to investigate gas vibrational relaxation rates and molecular energy transfer processes [23–26]. By the mid-20th century, however, the dominance of photoacoustic methods in gas analysis was supplanted by gas chromatography, owing to the latter's superior detection sensitivity.

The year 1960 marked a pivotal moment with the invention of the laser, which profoundly revolutionized the field of optics and, in turn, reinvigorated the research momentum behind the photoacoustic effect [27]. Lasers, characterized by their unique advantages—including high power, exceptional spectral purity, superior stability, and outstanding repeatability—provided an effective solution to the intrinsic limitations of conventional light sources in photoacoustic studies, such as inadequate energy output and suboptimal spectral properties.

The emergence of laser technology catalyzed the first major resurgence in photoacoustic gas analysis. In 1968, Kerr and Atwood pioneered the development of a laser-based absorption spectrophone for gas analysis [28]. Their experimental setup utilized a pulsed ruby laser and a modulated continuous-wave CO_2 laser as respective light sources, with acoustic signals being captured by a capacitive microphone. This approach enabled highly sensitive measurements of gas absorption characteristics. Equipped with distinct advantages such as ultrahigh sensitivity, excellent selectivity, and rapid response capabilities, laser-driven photoacoustic gas analysis technology demonstrated significant application potential across diverse fields including environmental monitoring, industrial process control, and atmospheric science research.

The advancement of laser technology in photoacoustic gas analysis coincided with a renewed scientific focus on the photoacoustic effect in solid materials. This resurgence can be largely credited to researchers at Bell Laboratories. Motivated by the remarkable sensitivity achieved in gas-phase photoacoustic detection, they initiated comprehensive investigations into the solid-state photoacoustic phenomenon in the early 1970s. In 1973, Rosencwaig successfully demonstrated photoacoustic spectra across a range of materials, extended the technique's applicability to the spectroscopic analysis of biological specimens, and contributed significantly to refining the theoretical understanding of the photoacoustic effect in solids [29, 30].

1.3 Photoacoustic Technology in Biomedical Applications

1.3.1 Photoacoustic Imaging Technology

In the 1980s, the photoacoustic effect began to progressively make its way into the biomedical field. Due to the strong optical scattering properties of biological tissues, conventional optical imaging techniques faced severe limitations in terms of penetration depth and spatial resolution. In contrast, photoacoustic imaging detects ultrasound waves, which experience significantly lower scattering and attenuation in biological tissues compared to light. This allows the technique to achieve deep-tissue imaging while preserving high resolution [31, 32]. This distinct advantage endowed the photoacoustic effect with tremendous application potential in biomedicine.

As early as 1964, Amar et al. conducted a pioneering experiment in which they observed laser pulse-induced elastic waves in the eyes of living rabbits for the first time. They employed a ruby laser to generate a pulse train with an average energy of 50 mJ and a pulse width of 1 microsecond, directing the laser onto the rabbits' retinas. A barium titanate-based ultrasound detector was placed on the rabbits' occipital bone, successfully capturing acoustic signals that had propagated through the brain and skull. The experiment confirmed that these acoustic signals corresponded to elastic waves within the ultrasonic frequency range, rather than shock waves, with a dominant frequency of approximately 40 kHz [33]. This study is widely regarded as the first attempt at biomedical photoacoustic imaging. Subsequently, researchers started exploring the application of the photoacoustic effect in the analysis of biological tissues. With the rapid advancement of laser technology, ultrasound detection methods, and data acquisition and processing systems, research on the photoacoustic effect entered a mature phase, and its focus fully shifted toward the biomedical field, sparking a revolution in biomedical imaging.

Since the late 1990s, photoacoustic imaging technology has entered a golden era of comprehensive development. Photoacoustic tomography (PAT) and photoacoustic microscopy (PAM) have emerged as the two core technical branches, continuously achieving breakthroughs in key performance metrics such as spatial resolution, penetration depth, and imaging speed [31].

Photoacoustic tomography (PAT) is primarily oriented toward deep-tissue imaging, achieving penetration depths of up to several centimeters (typically 1–5 cm in biological tissues) with submillimeter spatial resolution [34]. By comprehensively scanning the sample to collect photoacoustic signals from different positions, the technology reconstructs three-dimensional structural images using algorithms such as back-projection or iterative reconstruction. Furthermore, PAT can provide rich functional information: for example, based on the differences in absorption spectra of endogenous absorbers such as hemoglobin, oxyhemoglobin, and melanin, it enables quantitative measurement of physiological parameters such as tissue oxygen saturation and hemoglobin concentration. This offers important support for the early diagnosis of diseases. In contrast, photoacoustic microscopy (PAM) is mainly focused on microscopic imaging, achieving submicron-level spatial resolution but with a relatively shallow penetration depth (usually several hundred micrometers). By combining the high resolution of optical microscopy with the deep-penetration capability of ultrasound detection, PAM can visualize microstructures in biological tissues, such as capillary networks and cellular morphology [35].

Alongside the widespread adoption of photoacoustic imaging, its theoretical foundation has been continuously refined. Researchers have conducted more in-depth investigations into the physical mechanisms of the photoacoustic effect and established more precise mathematical models that account for complex factors such as tissue heterogeneity, optical scattering, and heat conduction. These advancements provide more reliable theoretical support for image reconstruction and quantitative analysis [36].

1.3.2 Laser-Generated Ultrasound Technology

In the evolutionary trajectory of photoacoustic imaging technology, researchers have made a critical observation: the bandwidth of photoacoustic signals generated via the photoacoustic effect is substantially broader than that of ultrasonic signals produced through the inverse piezoelectric effect [37]. This pivotal characteristic has unlocked new possibilities for the application of ultrasound technology in biomedicine—greater bandwidth enables higher temporal and spatial resolution, allowing for the visualization of finer structural details and dynamic changes within biological tissues, which holds irreplaceable value in contexts such as high-precision imaging and functional monitoring.

Building upon this foundational insight, LGUS technology has progressively matured. It inherits the core energy conversion logic of the photoacoustic effect, while achieving innovations in technical implementation. Unlike the passive mode of conventional photoacoustic imaging, which relies on the intrinsic optical absorption of biological tissues (e.g. endogenous absorbers such as hemoglobin and melanin), LGUS technology actively regulates the entire energy conversion process through the use of artificially engineered photoacoustic composite materials. By rationally designing the composition and structure of such materials, it becomes possible to precisely modulate key stages including optical absorption, heat conduction, and acoustic radiation, thereby achieving the generation of broadband, high-amplitude ultrasonic signals [38]. Furthermore, the inherent electrode-free design of LGUS provides natural immunity to electromagnetic interference (EMI), while its potential for miniaturization and flexibility—realizable through techniques such as thin-film deposition and optical fiber integration—makes it exceptionally well-suited to the varied demands of precision medicine for advanced ultrasound solutions [7].

In the realm of basic research, LGUS serves as a significant interdisciplinary platform bridging biomedical engineering, materials science, acoustics, and related disciplines. Its core conversion mechanism involves multiphysics-coupled processes such as optical absorption, thermal conduction, and elastic wave dynamics; in-depth investigation of this mechanism contributes to revealing fundamental principles of multiscale energy conversion. Simultaneously, the design and optimization of photoacoustic composite materials have stimulated technological advances in areas such as functional nanomaterials and polymer composites [38]. In biomedical applications, high-resolution imaging systems based on LGUS are anticipated to enable accurate detection of pathological structures such as vascular plaques [3]. Its capacity for precise ultrasonic modulation may also provide a single-neuron-level tool for neuroscience research, opening new pathways for the treatment of neurological disorders including Parkinson's disease and epilepsy [39]. Moreover, LGUS technology demonstrates considerable potential in fields such as minimally invasive surgery and targeted drug delivery, promising to drive the evolution of clinical diagnosis and treatment toward more precise, less invasive, and safer paradigms [6, 40, 41]. Therefore, advancing research on LGUS technology holds not only significant scientific and theoretical value but also notable potential for clinical translation and socioeconomic impact.

Subsequent chapters of this book will conduct systematic and in-depth thematic exploration of LGUS technology. First, the theoretical foundations and technical principles of LGUS will be comprehensively reviewed, summarizing key advances in physical modeling and mechanistic understanding. Next, focus will be placed on the structural designs and material systems of LGUS devices, highlighting pivotal directions for technological breakthroughs. Finally, practical application scenarios and implementation strategies of LGUS in domains such as biomedical imaging and precision therapy will be elaborated upon, with the aim of providing comprehensive technical support and academic reference to further academic investigation and industrial development of LGUS technology.

References

1 Zhou, Q., Lam, K.H., Zheng, H. et al. (2014). Piezoelectric films for high-frequency ultrasonic transducers in biomedical applications. *Progress in Materials Science*, 66: 87–111.

2 Engdahl, G. (ed.) (2000). *Handbook of giant magnetostrictive materials*. Academic Press.

3 Chen, S.L. (2016). Review of laser-generated ultrasound transmitters and their applications to all-optical ultrasound transducers and imaging. *Applied Sciences*, *7*(1): 25.

4 Bychkov, A., Simonova, V., Zarubin, V. et al. (2018). The progress in photoacoustic and laser ultrasonic tomographic imaging for biomedicine and industry: A review. *Applied Sciences*, *8*(10): 1931.

5 McDonald, F.A. (1980). Photoacoustic effect and the physics of waves. *American Journal of Physics, 48*(1): 41–47.

6 Baac, H.W., Ok, J.G., Park, H.J. et al. (2010). Carbon-nanotube optoacoustic lens for focused ultrasound generation and high-intensity targeted therapy. *Applied Physics Letters, 97*(23): 234104.

7 Kim, J., Kim, H., Chang, W.Y. et al. (2019). Candle-soot carbon nanoparticles in photoacoustics: advantages and challenges for laser ultrasound transmitters. *IEEE Nanotechnology Magazine, 13*(3): 13–28.

8 Manohar, S. and Razansky, D. (2016). Photoacoustics: a historical review. *Advances in Optics and Photonics, 8*(4): 586–617.

9 Smith, W. (1873). The action of light on selenium. *Journal of the Society of Telegraph Engineers*, 2: 31–33.

10 Mims, F.M. III, (1980). Alexander Graham Bell and the photophone: the centennial of the invention of light-wave communications, 1880–1980. *Optics News, 6*(1): 8–16.

11 Bell, A.G. (1880). On the production and reproduction of sound by light. *American Journal of Science*, 118: 305–324.

12 Bell, A.G. (1880). Selenium and the photophone. *Nature*, 22: 500–503.

13 Bell, A.G. (1881). Production of sound by radiant energy. *Journal of the Franklin Institute*, 111: 401–428.

14 Bell, A.G. (1881). The spectrophone. *Bulletin of the Philosophical Society*, 4: 42.

15 Mercadier, E. (1881). Sur la radiophonie. *Journal de Physique Théorique et Appliquée*, 10: 53–68.

16 Mercadier, E. (1881). Sur la radiophonie (2e mémoire). *Journal de Physique Théorique et Appliquée*, 10: 147–154.

17 Preece, W.H. (1880). On the conversion of radiant energy into sonorous vibrations. *Proceedings of the Royal Society of London*, 31: 506–520.

18 Röntgen, W.C. (1881). Ueber Töne, welche durch intermittirende Bestrahlung eines Gases entstehen. *Annalen der Physik*, 248: 155–159.

19 Tyndall, J. (1880). Action of an intermittent beam of radiant heat upon gaseous matter. *Proceedings of the Royal Society of London*, 31: 307–317.

20 Veingerov, M.L. (1938). A method of gas analysis based on the Tyndall-Röntgen optico-acoustic effect. *Doklady Akademii Nauk SSSR*, 19: 687–688.

21 Veingerov, M.L. (1946). An optical-acoustic method of gas analysis. *Nature*, 158: 28–29.

22 Luft, K.F. (1975). Infrared techniques for the measurement of carbon monoxide. *Annals of Occupational Hygiene*, 18: 45–51.

23 Veingerov, M.L. (1945). Spectrophone – an instrument for investigation of infrared absorption spectra of gases and for quantitative and

qualitative spectrum analysis of multi-component gas mixtures. *Doklady Akademii Nauk SSSR*, 46: 182.

24 Gorelik, G. (1946). On a possible method of studying the energy exchange time between the different degrees of freedom of molecules in a gas. *Doklady Akademii Nauk SSSR*, 54: 779.

25 Slobodskaya, P.V. (1948). Determination of the energy transfer rate from vibrational to translational molecular motion by means of a spectrophone. *Izvestiya Akademii Nauk SSSR*, 12: 656–662.

26 Cottrell, T.L. (1950). The absorption of interrupted infra-red radiation. *Transactions of the Faraday Society*, 46: 1025–1030.

27 Rosencwaig, A. (1980). *Photoacoustics and photoacoustic spectroscopy*. John Wiley & Sons.

28 Kerr, E.L. and Atwood, J.G. (1968). The laser illuminated absorptivity spectrophone: a method for measurement of weak absorptivity in gases at laser wavelengths. *Applied Optics*, *7*(5): 915–921.

29 Rosencwaig, A. (1975). Photoacoustic spectroscopy of solids. *Physics Today*, *28*(9): 23–30.

30 Harshbarger, W.R. and Robin, M.B. (1973). Opto-acoustic effect. Revival of an old technique for molecular spectroscopy. *Accounts of Chemical Research*, *6*(10): 329–334.

31 Beard, P. (2011). Biomedical photoacoustic imaging. *Interface Focus*, *1*(4): 602–631.

32 Xu, M. and Wang, L.V. (2006). Photoacoustic imaging in biomedicine. *Review of Scientific Instruments*, *77*(4).

33 Amar, L. (1964). Detection d'omes elastiques (ultrasonores) sur l'os occipital, induites par impulsions laser dans l'oeil d'un lapin. *CR Acad. Sc. Paris*, 259: 3653–3655.

34 Wang, L.V. and Yao, J. (2016). A practical guide to photoacoustic tomography in the life sciences. *Nature Methods*, *13*(8): 627–638.

35 Yao, J. and Wang, L.V. (2013). Photoacoustic microscopy. *Laser & Photonics Reviews*, *7*(5): 758–778.

36 Wang, L. (ed.) (2017). *Photoacoustic imaging and spectroscopy*. CRC Press.

37 Kruger, R.A. (1994). Photoacoustic ultrasound. *Medical Physics*, *21*(1): 127–131.

38 Lee, T., Baac, H.W., Li, Q., and Guo, L.J. (2018). Efficient photoacoustic conversion in optical nanomaterials and composites. *Advanced Optical Materials*, *6*(24): 1800491.

39 Du, Z., Chen, G., Li, Y. et al. (2024). Photoacoustic: a versatile nongenetic method for high-precision neuromodulation. *Accounts of Chemical Research*, *57*(11): 1595–1607.

40 Di, J., Kim, J., Hu, Q. et al. (2015). Spatiotemporal drug delivery using laser-generated-focused ultrasound system. *Journal of Controlled Release*, 220: 592–599.

41 Lee, T., Luo, W., Li, Q. et al. (2017). Laser-induced focused ultrasound for cavitation treatment: toward high-precision invisible sonic scalpel. *Small*, *13*(38): 1701555.

2

Laser-Generated Ultrasound Theory

2.1 Photo-Thermal-Acoustic Conversion Model

Laser-generated ultrasound (LGUS) arises from a three-stage process: light absorption, photothermal conversion, and thermoelastic expansion. When a pulsed laser irradiates a material, the absorbing medium takes in the optical energy, resulting in a rapid temperature rise. This induces transient thermal expansion of the material, followed by rapid contraction upon cooling which in turn generates ultrasonic waves. Consequently, LGUS signal generation can be modeled through thermodynamic energy relations and equations of state, coupled with heat conduction equations and equations for pressure fluctuations, enabling a rigorous derivation of the LGUS formation process [1]. In particular, the temperature field $T(\vec{r}, t)$ resulting from pulsed laser heating can be obtained by solving the heat conduction equation with an appropriate source term [2]:

$$\rho C_P \frac{\partial T(\vec{r}, t)}{\partial t} = \nabla(k \nabla T) + H(\vec{r}, t) \tag{2.1}$$

where ρ is the density (kg m^{-3}), k is the thermal conductivity (W $\cdot$ m^{-1} $\cdot$ K^{-1}), and $H(\vec{r}, t)$ is the nonradiative thermal field generated by light absorption (W m^{-3}). When the thermal constraint condition is satisfied, the thermal diffusion represented by the first term on the right-hand side of the equation can be ignored, $\nabla(k \nabla T) = 0$. The temperature field calculated by Eq. (2.1) can be substituted into the pressure wave equation to solve the pressure field $p(\vec{r}, t)$, and the equation form is expressed as follows:

$$\left[\nabla^2 - \frac{1}{c_s^2} \frac{\partial^2}{\partial t^2}\right] p(\vec{r}, t) = -\rho\beta \frac{\partial^2 T}{\partial t^2} \tag{2.2}$$

Note that the sound source modeled by the thermoelastic mechanism is represented on the right-hand side of the equation. The relationship between the pressure field and the thermal field can be obtained by combining Eqs. (2.1) and (2.2):

$$\left[\nabla^2 - \frac{1}{c_s^2}\frac{\partial^2}{\partial t^2}\right] p(\vec{r}, t) = -\frac{\beta}{C_P}\frac{\partial H(\vec{r}, t)}{\partial t} \tag{2.3}$$

For analytical simplicity, the one-dimensional form of Eq. (2.3) may be considered:

$$\left[\frac{\partial^2}{\partial z^2} - \frac{1}{c_s^2}\frac{\partial^2}{\partial t^2}\right] p(z, t) = -\frac{\beta}{C_P}\frac{\partial H(z, t)}{\partial t} \tag{2.4}$$

The analytical solution to Eq. (2.4) can be obtained via the Green's function method [3]. In the one-dimensional case, the solution takes the following form:

$$p(z, t) = \frac{\beta c_s}{C_P}\int dz' H\left(z', t - \frac{|z - z'|}{c_s}\right) \tag{2.5}$$

To generate strong LGUS signals, the following two conditions need to be satisfied [4].

i) **Thermal confinement:** The thermal energy must be deposited faster than it diffuses away, ensuring a maximum localized temperature rise.
ii) **Stress confinement:** The laser heating must occur on a time scale shorter than the thermal expansion time of the material; that is, heating should take place at approximately constant volume, allowing high thermoelastic pressure to build up.

These two conditions can also be expressed as: $\tau_l < \tau_{th}$ and $\tau_l < \tau_s$, where τ_l represents the pulse width of the pulsed laser, referring to the full width at half maximum (FWHM) of the laser power peak. τ_{th} is the thermal diffusion time, and τ_s is the stress relaxation time [5, 6]. τ_{th} and τ_s can be expressed by using the following equations:

$$\tau_{th} = \frac{1}{\mu_a^2 \times \alpha_{th}} \tag{2.6}$$

$$\tau_s = \frac{1}{\mu_a \times c_s} \tag{2.7}$$

Among these parameters, μ_a is the light absorption coefficient (m^{-1}), which defines the energy density of light absorption per unit length. Here, $\mu_a = \ln 10 \cdot A/l \approx 2.3\,A/l$, where A is the absorbance (dimensionless) defined as $A = 20\,\lg(I_i/I_o)$; I_i and I_o are the incident and outgoing light intensities, respectively; l is the material thickness (m); c_s is the sound velocity in the material ($\mathrm{m\,s^{-1}}$); and α_{th} is the thermal diffusion coefficient ($\mathrm{m^2\,s^{-1}}$). Consequently, to generate an LGUS signal, the laser pulse width must be as short as possible for a given material. Therefore, a nanosecond pulsed laser is typically employed as the excitation light source. Under the conditions described above, the amplitude (P_0) of the LGUS signal can be expressed as [3]:

$$P_0 = \Gamma \cdot A \cdot \frac{F}{L} \tag{2.8}$$

where $\Gamma = \frac{\beta c_s^2}{C_P}$ is the Grüneisen parameter, β is the coefficient of thermal expansion ($\mathrm{K^{-1}}$), C_P is the constant pressure heat capacity ($\mathrm{J \cdot K^{-1} \cdot Kg^{-1}}$), F is the laser energy density ($\mathrm{J\,m^{-2}}$), and L is the characteristic length of the absorption region (m) defined as $L = c_s \tau_l + 1/\mu_a$.

As shown in Eq. (2.8), the amplitude of the LGUS signal is directly proportional to the Grüneisen parameter, the absorbance, and the laser energy density of the material, and inversely proportional to the characteristic length. Therefore, in order to obtain strong LGUS signals under a fixed laser energy density, materials need to have a high optical absorption coefficient, a high thermal expansion coefficients, a high sound velocity, and a low heat capacity. However, it is challenging for a single material to simultaneously satisfy all these requirements. As a result, composite materials—often combining strongly absorbing constituents with materials exhibiting high thermoelastic response—are typically employed to achieve efficient photoacoustic conversion.

For photoacoustic composite materials, the amplitude of the resulting LGUS signal originates from combined contributions of two layers, and the total output sound pressure can be expressed as $P_0 = P_A + P_P$. Due to the low thermal expansion coefficient of the light-absorbing layer (e.g. metal thin films or carbon nanomaterials), its contribution to the LGUS signal amplitude is often negligible ($P_A/P_P \ll 1$). Consequently, the resulting LGUS signal is mainly provided by the thermal expansion layer, $P_0 \cong P_P$. Therefore, in order to obtain high-amplitude LGUS signals, in addition to the influencing factors in Eq. (2.8), improving the heat transfer between the light-absorbing material and the thermal expansion material to enhance energy conversion efficiency is also a critical consideration.

To explain the heat transfer process in such composite structure, a layered model is established, which consists of a thin light-absorbing layer and a thick thermal expansion layer [3]. The light-absorbing thin layer absorbs incident light and converts it into heat, with a portion of the thermal energy being transferred to the adjacent thermal expansion layer. According to the conservation of energy, the proportion fraction of thermal energy (γ = thermal energy per layer/absorbed light energy) is expressed as:

$$\frac{F_A}{A \cdot F} + \frac{F_P}{A \cdot F} = \gamma_A + \gamma_P = 1 \tag{2.9}$$

The subscripts A and P denote the light-absorbing layer and the thermal expansion layer, respectively. The amplitude of the output LGUS signal is determined by the combination of the LGUS signal amplitudes generated by each layer. Using Eqs. (2.8) and (2.9), the amplitude of the output LGUS signal can be expressed as:

$$P_0 = P_A + P_P \cong P_P = A \cdot F \cdot \gamma_P \left(\frac{\Gamma}{L}\right)_P \tag{2.10}$$

In this case, in order to minimize the thermal energy loss ($A \cdot F \cdot \gamma_A$) within the light-absorbing layer, the ratio of γ_A and γ_P should be reduced to promote heat transfer and thus increase the amplitude of the LGUS signal. The value of γ_P can be obtained by solving the heat transfer equation, and the fraction of thermal energy transferred to the thermal expansion layer can be estimated as [3]:

$$\gamma_P = \frac{1}{\frac{C_A}{C_P} + 1} \tag{2.11}$$

$$\frac{C_A}{C_P} = \frac{d_A \cdot S \cdot (\rho C_P)_A}{l_{th,P} \cdot S \cdot (\rho C_P)_P} = \frac{d_A (\rho C_P)_A}{l_{th,P} (\rho C_P)_P} \tag{2.12}$$

In the formula, C_A/C_P is the ratio of the heat capacity of the light-absorbing layer and the thermal expansion layer. Here, $l_{th,P} = \sqrt{1.269\alpha_{th}\tau_l}$ is the thermal penetration depth of the thermal expansion layer, where α_{th} is the thermal diffusion coefficient, and d_A is the thickness of the absorbing layer. From Eqs. (2.11) and (2.12), it can be seen that γ_P can be increased by either reducing the specific heat capacity of the light-absorbing layer (e.g. by decreasing d_A) or increasing the specific heat capacity of the thermal expansion layer (e.g. by increasing $l_{th,P}$).

2.2 Laser-Induced Damage Threshold Theory

Generally, the generation of high-amplitude ultrasonic signals necessitates the irradiation of photoacoustic materials with high-energy pulsed lasers [2, 7]. However, the use of such high-energy pulsed lasers carries the risk of inducing damage to the material [8]. The laser-induced damage is closely related to the thickness of the photoacoustic composites, as well as their structural details [9]. Consequently, advancing laser-generated ultrasound (LGUS) technology necessitates a thorough investigation into the underlying mechanisms of laser-induced damage and the development of a robust theoretical model to predict it.

Given that the thermal penetration depth and materials ablation depth induced by a nanosecond pulsed laser are much smaller than the laser irradiation area, a one-dimensional heat conduction model is justified for the entire process. Before the surface temperature reaches the melting point or pyrolysis temperature of the material, the heat conduction equation and the associated boundary conditions at any interior point inside the material can be expressed as [10]:

$$\frac{\partial^2 T_S(x,t)}{\partial x^2} = \frac{1}{\alpha_S} \cdot \frac{\partial T_S(x,t)}{\partial t}, \quad 0 \le x < +\infty \tag{2.13}$$

$$-\lambda_S \frac{\partial T_S(x,t)}{\partial x} = \mu A_S I, \quad x = 0 \tag{2.14}$$

$$T_S(x,t) = T_0, \quad x \to +\infty \tag{2.15}$$

$$T_S(x,t) = T_0, \quad t = 0 \tag{2.16}$$

where T_S, α_S, λ_S, A_S, and μ are temperature, thermal diffusivity, thermal conductivity, light absorption, and photothermal conversion efficiency, respectively. T_0 is the initial ambient temperature, and I is the incident laser power density. Assuming that a value of $T_S(x, t)$ satisfies the boundary conditions (2.15) and (2.16), its general expression can be formulated as:

$$T_S(x,t) = T_E(t)e^{-\frac{x}{\omega(t)}} + T_0 \tag{2.17}$$

where $T_E(t)$ is a function of time only, and ω(t) is the depth of thermal penetration. The partial derivative of the function expressed in Eq. (2.17) can be written as:

$$\frac{\partial T_S(x,t)}{\partial x} = -\frac{1}{\omega(t)} T_E(t)e^{-\frac{x}{\omega(t)}} \tag{2.18}$$

$$\frac{\partial^2 T_S(x,t)}{\partial x^2} = \frac{1}{\omega^2(t)} T_E(t)e^{-\frac{x}{\omega(t)}} \tag{2.19}$$

$$\frac{\partial^2 T_S(x,t)}{\partial t} = \frac{dT_E(t)}{dt} e^{-\frac{x}{\omega(t)}} + \frac{x}{\omega^2(t)} \frac{d\omega(t)}{dt} T_E(t)e^{-\frac{x}{\omega(t)}} \tag{2.20}$$

The substitution of Eqs. (2.19) and (2.20) into Eq. (2.13), and of Eq. (2.18) into Eq. (2.14), results in

$$\frac{dT_E(t)}{dt} = \frac{1}{\omega^2(t)}\left(\alpha_S - x\frac{d\omega(t)}{dt}\right)T_E(t) \tag{2.21}$$

$$T_E(t)e^{-\frac{x}{\omega(t)}} = \frac{\mu A_S I}{\lambda_S}\omega(t) \tag{2.22}$$

When $x = 0$, Eqs. (2.21) and (2.22) can be transformed into

$$\frac{dT_E(t)}{dt} = \frac{\alpha_S}{\omega^2(t)}T_E(t) \tag{2.23}$$

$$T_E(t) = \frac{\mu A_S I}{\lambda_S}\omega(t) \tag{2.24}$$

Derivation of Eq. (2.24) gives

$$\frac{dT_E(t)}{dt} = \frac{\mu A_S I}{\lambda_S}\frac{d\omega(t)}{dt} \tag{2.25}$$

And the thermal penetration depth $\omega(t)$ satisfies the boundary conditions:

$$\omega(t) = 0, t = 0 \tag{2.26}$$

Substituting Eqs. (2.25) and (2.24) into (2.23), along with the application of boundary condition (2.26), yields:

$$\omega(t) = \sqrt{2\alpha_S t} \tag{2.27}$$

The heat diffusion coefficient is defined as $\alpha = \frac{\lambda}{\rho C}$. Inserting Eq. (2.27) into Eq. (2.24) gives:

$$T_E(t) = \mu A_S I\sqrt{\frac{2t}{C_S\rho_S\lambda_S}} \tag{2.28}$$

The substitution of Eqs. (2.28) and (2.27) into Eq. (2.17) results in:

$$T_S(x,t) = \mu A_S I\sqrt{\frac{2t}{C_S\rho_S\lambda_S}}e^{-\frac{x}{\sqrt{2\alpha_S t}}} + T_0 \tag{2.29}$$

When $x = 0$, the surface temperature T_W(t) of the material is

$$T_W(t) = T_S(0,t) = \mu A_S I\sqrt{\frac{2t}{C_S\rho_S\lambda_S}} + T_0 \tag{2.30}$$

Under nanosecond pulse laser irradiation, the temperature rise within photoacoustic composite material is not primarily governed by free-electron heating, due to their high absorption rate and low reflectivity. Instead, it arises mainly from heat generated through chemical bond breaking or lattice thermal vibration. Once the laser energy exceeds a critical threshold, damage occurs predominantly when the surface temperature exceeds either the melting point of the light-absorbing material or the pyrolysis temperature of the polymer. This leads to ablative processes in the irradiation area, including processes such as polymer pyrolysis and oxidation of the light-absorbing material. When irradiated with a multi-pulse laser, the surface temperature of the material accumulates with the increase of the number of pulses. Assuming that the material remains undamaged at the (n–1)th pulse and is damaged at the end of the nth pulse, that is, $t = nt_v + \tau$, where $t_v = 1/v$ is the pulse repetition period, v is the pulse repetition frequency, and τ is the pulse width. Further assuming that the thermal convection process inside the material ceases before the next pulse arrives, the temperature of the material is equal throughout the thermal diffusion depth. The material is evenly divided into N thin layers in the depth x direction, with each layer having a thickness of $\omega(t_v)$. Within the first pulse time τ, materials within the thermal diffusion depth $\omega(\tau)$ reaches a uniform temperature $T_W(\tau)$, given by:

$$T_W(\tau) = \mu A_S I \sqrt{\frac{2\tau}{C_S \rho_S \lambda_S}} + T_0 \tag{2.31}$$

$$\omega(\tau) = \sqrt{2\alpha_S \tau} \tag{2.32}$$

Within time t_v, the thermal diffusion depth $\omega(t_v)$ inside the material is

$$\omega(t_v) = \sqrt{2\alpha t_v} \tag{2.33}$$

From the heat exchange equation $V_1 T_1 = V_2 T_2$, the temperature within the material depth $x \leq \omega(t_v)$ is derived as:

$$T_1 = T_W(\tau) \cdot \frac{\omega(\tau)}{\omega(t_v)} + T_0 \cdot \frac{\omega(t_v) - \omega(\tau)}{\omega(t_v)} \tag{2.34}$$

When the second laser pulse ends, the time is $t = t_v + \tau$

$$T_W(t_v + \tau) = \mu A_S I \sqrt{\frac{2\tau}{C_S \rho_S \lambda_S}} + T_1 \tag{2.35}$$

When $t = 2t_v$, under the assumption that the heat exchange between the material within the depth $\omega(t_v)$ and the second thin layer is complete, the final temperature of the first thin layer can be obtained as:

$$T_2 = T_W(t_v + \tau) \cdot \frac{\omega(\tau)}{\omega(t_v)} + \frac{1}{2}(T_0 + T_1) \cdot \frac{\omega(t_v) - \omega(\tau)}{\omega(t_v)} \tag{2.36}$$

By using the impulse theorem method to analytically solve the heat conduction equation, at the n–2th pulse $t = (n-2)t_v + \tau$

$$T_W((n-1)t_v + \tau) = \mu A_S I \sqrt{\frac{2\tau}{C_S \rho_S \lambda_S}} + T_{n-1} \tag{2.37}$$

The temperature of the first thin layer before the nth pulse ($t = (n-1)t_v$) can be obtained as:

$$T_{n-1} = T_W((n-1)t_v + \tau) \cdot \frac{\omega(\tau)}{\omega(t_v)} + \frac{1}{2}(T_{n-2} + T_{n-3}) \cdot \frac{\omega(t_v) - \omega(\tau)}{\omega(t_v)} \tag{2.38}$$

When the melting point or pyrolysis temperature T_m of the material is reached after the nth pulse

$$T = T_W((n-1)t_v + \tau) = \mu A_S I \sqrt{\frac{2\tau}{C_S \rho_S \lambda_S}} + T_{n-1} = T_m \tag{2.39}$$

By solving the simultaneous Eqs. (2.35–2.39) using a computer inversion algorithm, the material incident laser power density threshold I_{th} and laser damage threshold F_{th} for a pulse number of n can be obtained as [11]:

$$I_{\text{th}} = \frac{81}{147 + 20(n-4)} \sqrt{\frac{\rho C \lambda}{\tau_l}} \frac{T_m - T_0}{\mu A_S}, \quad n \geq 5 \tag{2.40}$$

$$F_{\text{th}} = I_{\text{th}} \cdot n \cdot \tau_l = \frac{81n}{147 + 20(n-4)} \frac{\sqrt{\rho C \lambda \tau_l}(T_m - T_0)}{\mu A_S}, \quad n \geq 5 \tag{2.41}$$

where n is the number of laser pulses, λ is the thermal conductivity of the composite material ($\text{W} \cdot \text{m}^{-1} \cdot \text{K}^{-1}$), τ_l is the width of the incident laser pulse (ns), C is the specific heat capacity of the composite material ($\text{J} \cdot \text{kg}^{-1} \cdot \text{K}^{-1}$), and ρ is the density of the composite material ($\text{kg} \cdot \text{m}^{-3}$).

Eq. (2.41) indicates that when the number of laser pulses, laser pulse width, and spot area are held constant, the laser damage threshold of photoacoustic composite materials is chiefly determined by their density, thermal conductivity, and specific heat capacity.

2.3 Frequency Characteristic of Laser-Generated Ultrasound

In practical application ultrasound applications, frequency characteristics emerge as a critical parameter that fundamentally dictates the trade-off between resolution and penetration depth—a dynamic balance that directly governs imaging quality and diagnostic accuracy [12]. Therefore, advancing laser-generated ultrasound (LGUS) technology necessitates a more rigorous theoretical derivation and mechanistic analysis of the signal excitation process.

Expressing the thermal field function in Eq. (2.5) as $H(z, t) = I_0 f(t)g(z)$, where I_0 denotes the peak intensity (W m^{-2}), while $f(t)$ and $g(z)$ represent the temporal heating function and the spatial light absorption function, respectively. Considering the acoustic field propagating along the positive direction of the z-axis and neglecting the influence of other coefficients, the resulting LGUS signal at the material surface can be estimated as $p(t)\sim f * g$. A detailed explanation of this relationship will be provided subsequently.

In order to elucidate the formation process of the LGUS pulse, a one-dimensional model is established. In this model, it is assumed that a laser pulse characterized by a Gaussian time function profile $f(t)$ is irradiated onto the photoacoustic composite. The incident optical energy is considered to be fully absorbed, and the spatial light absorption profile function $g(z)$ of the composite is treated as spatially homogeneous (regarded as a rectangular function, with a depth of $1/\mu_a$) [2, 13], i.e.

$$f(t) = I_0 exp\left[-\left(\frac{t}{2\sigma^2}\right)^2\right] \tag{2.42}$$

$$g(z) = g_0\left(H(0) - H\left(z - \frac{1}{\mu_a}\right)\right) \tag{2.43}$$

Here, $\sigma = \tau_l/2\sqrt{2ln2}$, g_0 is the light absorption amplitude and $H(z)$ denotes the Heaviside function.

$$H(z - z_0) = \begin{cases} 0\ (z < z_0) \\ 1\ (z > z_0) \end{cases} \tag{2.44}$$

The absorbed laser fluence is a certain value $F = g_0/\mu_a$. This relationship indicates that for composites with greater optical thickness (i.e., larger $1/\mu_a$), the light absorption amplitude g_0 must be smaller to maintain the same fluence F, and conversely, for thinner composites (i.e., smaller$1/\mu_a$), the light absorption amplitude g_0 is larger.

At this time, the thermal constraints are still satisfied, meaning that thermal diffusion during the laser pulse is negligible. Consequently, the optical absorption depth coincides with the depth of the heat source. Splitting the photoacoustic composite into numerous thin slices, the resulting LGUS signal can be treated as the sum of the signals generated by countless thin slices. Since each slice emits an acoustic wave with a temporal profile identical to $f(t)$, the overall photoacoustic signal can be understood as a convolution integral of the laser pulse temporal heating function $f(t)$ and the spatial light absorption function $g(z)$ as illustrated in Figure 2.1. Within this one-dimensional approximation, the photoacoustic signal can

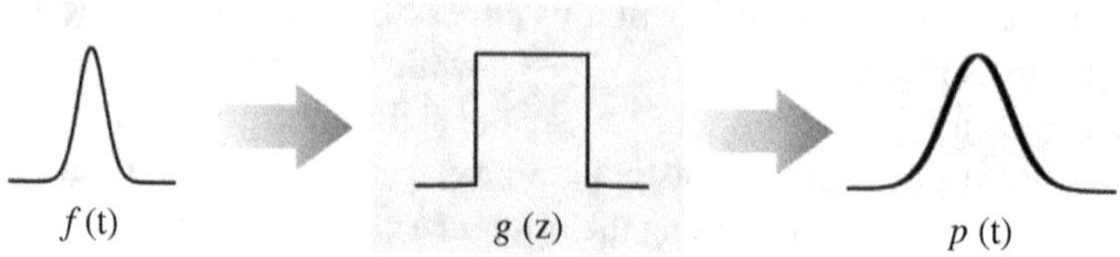

Figure 2.1 Schematic diagram of LGUS signal convolution [2].

be further estimated from Eq. (2.5) as:

$$p(t) \sim f * g = \int f(c_s t - z) g(z) dz \tag{2.45}$$

Based on the convolution relation, the pulse width τ_u of the LGUS signal can be viewed as the sum of the laser pulse width τ_l (associated with $f(t)$) and the time scale a (associated with g(z)). Consequently, and in combination with Eq. (2.7), τ_u can be expressed as the sum of the laser pulse width τ_l and the stress relaxation time τ_s

$$\tau_u = \tau_l + \frac{1}{c_s \mu_a} = \tau_l + \tau_s \tag{2.46}$$

And the length scale corresponding to the pulse width τ_u of the LGUS signal is given by

$$L = c_s \tau_u = c_s(\tau_l + \tau_s) = c_s \tau_l + \frac{1}{\mu_a} \tag{2.47}$$

It corresponds to the characteristic length L of the absorbing region in Eq. (2.8).

Based on the preceding derivation, the theoretical pulse width τ_u of the LGUS signal is obtained; however, this waveform can broaden due to effects such as inhomogeneous light absorption and non-instantaneous heat transfer. Consequently, the τ_u given by Eq. (2.46) represents the ideal, minimum achievable pulse width. Therefore, the corresponding upper bandwidth limit of the corresponding LGUS signal is

$$f_{\text{BWmax}} = \frac{1}{2\tau_u} = \frac{1}{2} \cdot \frac{1}{\tau_l + \tau_s} = \frac{1}{2} \cdot \frac{1}{\tau_l + 1/c_s \mu_a} \tag{2.48}$$

In the theoretical model, both the laser pulse width τ_l and the resulting LGUS signal pulse width τ_u are defined as full width at half maxima (FWHM). For a Gaussian-type pulse, the time-bandwidth product is about 0.5, which leads to the expression for the upper bandwidth of the LGUS signal is $f_{\text{BWmax}} = 1/2\tau_u$. It can be seen from Eq. (2.48) that the upper bandwidth of the LGUS signal is mainly determined by the laser pulse width, the material sound velocity, and the optical absorption coefficient.

Given that LGUS signals are typically generated using a nanosecond laser, τ_l is generally on the nanosecond scale. In case where the photoacoustic composite thickness l is relatively large (i.e., when μ_a is small, and thus τ_s is large), τ_l in Eq. (2.48) can be neglected if $\tau_l \ll \tau_s$ is satisfied. Under this condition,

$$f_{\text{BWmax}} \cong \frac{1}{2\tau_s} = \frac{c_s\mu_a}{2} \ (\tau_l \ll \tau_s) \tag{2.49}$$

This regime is referred as the short-pulse condition. The upper limit of the bandwidth of LGUS signal is primarily governed by the stress relaxation time τ_s, which is also determined by the sound velocity and light absorption coefficient of the material. To achieve a higher signal frequency, a reduction in τ_s is required, which can be accomplished by increasing either the material's optical absorption coefficient or its sound velocity. However, it should be noted that as τ_s decreases, its value will gradually approach τ_l, thus no longer satisfying the short pulse condition $\tau_l \ll \tau_s$. Under such circumstances, τ_l can no longer be neglected, and the upper limit of the signal bandwidth remains governed by the combined effect of both timescales as expressed in Eq. (2.48). Therefore, when estimating the frequency band of the LGUS signal, it is necessary to pay attention to the order-of-magnitude relationship between τ_l and τ_s, and select the appropriate formula from Eqs. (2.48) or (2.49) for calculation.

References

1 Lai, H.M. and Young, K. (1982). Theory of the pulsed optoacoustic technique. *The Journal of the Acoustical Society of America*, *72*(6): 2000–2007.

2 Lee, T., Baac, H.W., Li, Q., and Guo, L.J. (2018). Efficient photoacoustic conversion in optical nanomaterials and composites. *Advanced Optical Materials*, *6*(24): 1800491.

3 Xu, M. and Wang, L.V. (2006). Photoacoustic imaging in biomedicine. *Review of Scientific Instruments*, *77*(4): 041101.

4 Silva, A.D., Henriques, C.A., Malva, D.V. et al. (2020). Photoacoustic generation of intense and broadband ultrasound pulses with functionalized carbon nanotubes. *Nanoscale*, *12*(40): 20831–20839.

5 Vella, D., Pereira, D.A., Mrzel, A. et al. (2024). Picosecond photoacoustic generation of ultrasounds with composites of graphene-decorated gold nanoparticles. *Nano Energy*, 131: 110236.

6 Noimark, S., Colchester, R.J., Poduval, R.K. et al. (2018). Polydimethylsiloxane composites for optical ultrasound generation and multimodality imaging. *Advanced Functional Materials*, *28*(9): 1704919.

7 Baac, H.W., Ok, J.G., Lee, T., and Guo, L.J. (2015). Nano-structural characteristics of carbon nanotube–polymer composite films for high-amplitude optoacoustic generation. *Nanoscale*, *7*(34): 14460–14468.

8 Zhang, T., Yuan, J., Li, J. et al. (2024). Design and prediction of laser-induced damage threshold of CNT–PDMS optoacoustic transducer. *Ultrasonics*, 142: 107377.

9 Faraz, M., Abbasi, M.A., and Sang, P. (2020). Stretchable and robust candle-soot nanoparticle-polydimethylsiloxane composite films for laser-ultrasound transmitters. *Micromachines*, 11: 631.

10 Shen, Z.H., Zhang, S.Y., Lu, J., and Ni, X.W. (2001). Mathematical modeling of laser induced heating and melting in solids. *Optics & Laser Technology*, *33*(8): 533–537.

11 Torrisi, L., Borrielli, A., and Margarone, D. (2007). Study on the ablation threshold induced by pulsed lasers at different wavelengths. *Nuclear Instruments and Methods in Physics Research Section B: Beam Interactions with Materials and Atoms*, *255*(2): 373–379.

12 Kossoff, G. (2000). Basic physics and imaging characteristics of ultrasound. *World Journal of Surgery*, *24*(2): 134–142.

13 Huang, W., Chang, W.Y., Kim, J. et al. (2016). A novel laser ultrasound transducer using candle soot carbon nanoparticles. *IEEE Transactions on Nanotechnology*, *15*(3): 395–401.

3

Laser-Generated Ultrasound Device

The basic structure of a laser-generated ultrasound (LGUS) device comprises a photoacoustic material and a substrate, as shown in Figure 3.1. The photoacoustic material, which serves as the core functional component, can efficiently convert light energy into acoustic energy and is typically composed of a light-absorbing material and a thermal expansion material. The working principle, as shown in Figure 3.2, is that the light-absorbing material absorbs light energy and completes the conversion of light and heat, while the thermal expansion material makes the material expand periodically based on the principle of thermoelasticity, which then emits the acoustic wave [1].

Generally, LGUS device is planar one, as shown in Figure 3.3(a). It features a flat, broad emitting surface, often constructed with a uniform composite layer of light-absorbing material and thermal expansion material. When a pulsed laser beam is directed onto this planar surface, the rapid absorption of laser energy generates intense thermal stress within the thermal expansion material, causing it to expand and contract rapidly, thereby producing broadband ultrasound waves that propagate outward in all directions from the planar area. This kind of device produces a relatively wide ultrasound beam with lower spatial resolution [2]. In order to make the LGUS beam narrow, the development of focused device is proposed, where the composite layer is deposited in a curved or shaped transparent substrate, such as a concave lens, as shown in Figure 3.3(b) [3]. The geometry focusing nature of such device creates a more concentrated and directional ultrasound beam with enhanced spatial resolution.

Another widely used type of LGUS device is the fiber-optic LGUS device, as shown in Figure 3.4, whose development stems from the pursuit of higher imaging precision and smaller probe sizes in ultrasonic technology [4]. Leveraging the facile fabrication characteristic of photoacoustic materials, these materials can be directly integrated onto the end faces of optical

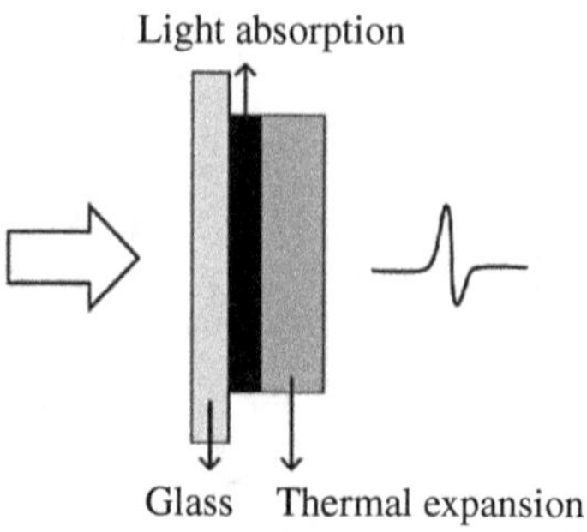

Figure 3.1 LGUS device with a basic three-layer structure.

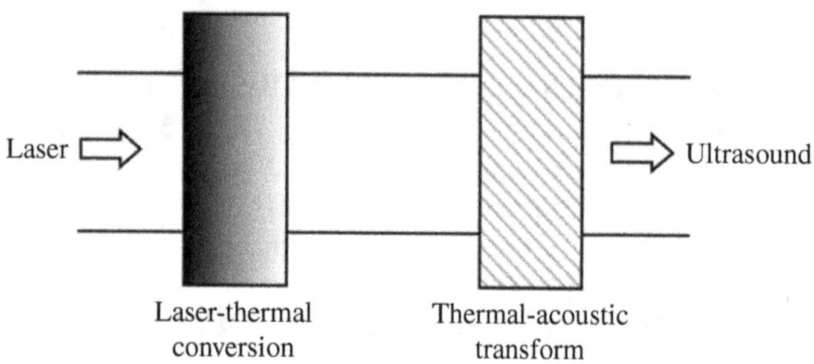

Figure 3.2 Schematic diagram of light-to-heat and heat-to-acoustic conversion.

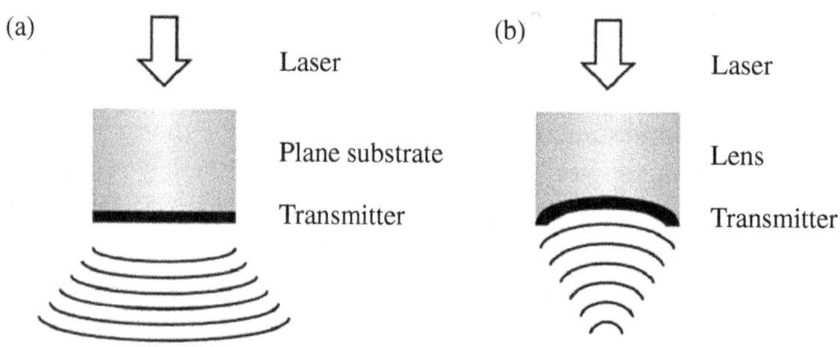

Figure 3.3 Schematic diagrams of planar (a) and focused (b) LGUS devices.

fibers, thereby enabling the manufacturing of fiber-optic LGUS devices. This innovative design not only addresses the miniaturization challenge of traditional ultrasonic probes but also provides a more accurate tool for clinical scenarios such as intravascular ultrasound imaging and ultrasonic neural stimulation, further expanding the application boundaries of LGUS technology.

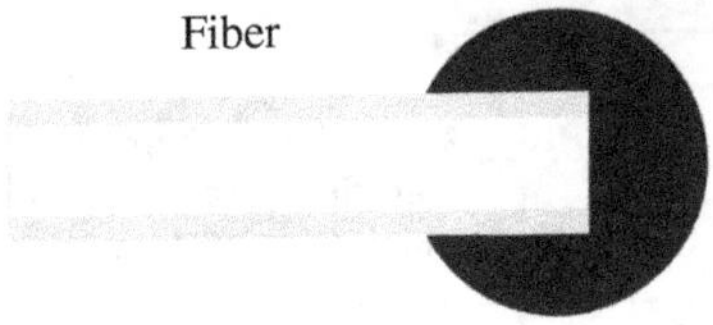

Figure 3.4 Schematic structure of a fiber-optic type LGUS device.

Enhancing the performance of photoacoustic materials is crucial for advancing the application of LGUS devices. The key properties of photoacoustic materials include light absorption coefficient, thermal conductivity, specific heat capacity, thermal expansion coefficient, damage threshold. In addition, other parameters such as acoustic impedance and material microstructure, which are crucial for the sound wave transmission efficiency, can also influence the performance of photoacoustic materials.

This chapter will cover the relevant content regarding photoacoustic materials—key components of LGUS devices. First, light-absorbing materials are introduced, including metal materials, carbon nanomaterials, and other related materials; subsequently, thermal expansion materials are presented, with a focus on analyzing materials such as polydimethylsiloxane (PDMS).

3.1 Light-absorbing Material

Light-absorbing materials play a paramount role in LGUS devices. Their primary function is to efficiently absorb light energy and convert it into heat. The efficacy of these materials is dictated by intrinsic properties including the optical absorption coefficient, thermal conductivity, and specific heat capacity.

The optical absorption coefficient describes the ability of a material to absorb light of a specific wavelength. It is defined as the rate of exponential attenuation of light intensity per unit distance, as described by the equation $I(x) = I_0 e^{-\alpha x}$, where $I(x)$ is the intensity of the light at depth x, I_0 is the incident light intensity, and α is the absorption coefficient. A higher absorption coefficient implies that the material is capable of absorbing a large amount of light energy over a short optical range. Thermal conductivity, expressed in $W \cdot m^{-1} \cdot K^{-1}$, measures a material's ability to conduct heat. It is defined as the heat flux per unit area under a unit temperature gradient. Specific heat capacity is the amount of heat required to raise a unit mass of a material by a

Table 3.1 Performance summary of representative light-absorbing materials.

Materials	Absorption coefficient (cm^{-1})	Thermal conductivity ($W \cdot m^{-1} \cdot K^{-1}$)	Specific heat capacity ($J\ Kg^{-1} \cdot K^{-1}$)	Thermal diffusion coefficient ($m^2 \cdot s^{-1}$)	References
CB	10^4–10^5	~0.03	~675	~10^{-5}	[5]
CNF		~1950	~750	~10^{-5}	[6]
SWCNTs	10^4–10^5	~3500	~600	~5.3×10^{-5}	[7–9]
MNCNTs	10^4–10^5	~700	~300	~10^{-5}	[8–10]
CSNPs			~750		[11]
Graphite		~350	~400		[10]
Graphene		~2000	~200		[10]
rGO		~300			[12]
$MAPbI_3$	10^4–10^5		~300	~1.45×10^{-7}	[11]

All are approximate values at room temperature.

unit temperature. Specific heat capacity and thermal conductivity together determine the thermal diffusivity, which measures the rate at which heat diffuses through a material. Table 3.1 shows the light absorption coefficient, thermal conductivity, specific heat capacity, and thermal diffusivity data for common light-absorbing materials.

An effective light-absorbing material should exhibit a high optical absorption coefficient, low specific heat capacity, high thermal conductivity, and an appropriate thermal diffusivity. A high absorption coefficient improves photothermal conversion efficiency, while a low specific heat capacity allows the material to achieve a larger temperature rise under the same incident light energy, thereby enhancing the output acoustic pressure of LGUS devices. High thermal conductivity promotes efficient heat transfer between the light-absorbing layer and the thermal expansion medium, enabling more heat to be delivered within a given timeframe. This contributes to higher output acoustic pressure and operating frequency. Thermal diffusivity primarily governs the rate of heat transfer between the light-absorbing material and the thermal expansion medium. A low thermal diffusivity can lead to poor heat dissipation, causing heat to accumulate within the light-absorbing layer. Excessive heat buildup may result in thermal damage to the device, lowering its laser-induced damage threshold. Therefore, selecting materials with high thermal diffusivity is critical for improving device durability and performance stability.

3.1.1 Metal-PDMS Devices

Metals, which are abundantly available in nature, exhibit excellent electrical and thermal conductivity and are widely utilized in modern industrial applications. They can be processed into thin films through various fabrication methods, such as sputtering and evaporation. With in-depth research on light absorption, photothermal effects, and sound wave generation, as well as the gradual refinement of the theory of photoacoustic effects in solid materials, metals have become the first choice for light-absorbing materials applied in LGUS devices due to their excellent thermal conductivity and high light absorption capacity.

When a pulsed laser irradiates a metal film, the film absorbs light and converts it into heat, successfully generating acoustic waves at MHz frequencies. For example, depositing a chromium (Cr) film onto a glass surface via sputtering allows the generation of ultrasonic signals with a bandwidth exceeding 100 MHz [13], but the signal amplitude is weak. To address the limitation of low acoustic pressure output from pure metal films, researchers have developed and extensively studied metal-based composites. They typically combine metals such as gold (Au), chromium (Cr), or germanium (Ge) with PDMS, leveraging the efficient light-to-heat conversion of the metal layer and the pronounced thermal expansion of PDMS to significantly improve ultrasonic generation. The composite structures produce an ultrasonic output with enhanced amplitude. Specifically, the output exhibits a broad bandwidth (~180% at −6 dB) and operates at a high frequency of ~80 MHz [14].

Although metal-film-based LGUS devices are capable of generating broadband ultrasound, their performance is constrained by strong optical reflection. To enhance optical absorption efficiency, one viable solution is to optimize the metal film thickness and material composition. Furthermore, the introduction of surface nanostructures or optical enhancement coatings can reduce reflection losses. For example, one approach involves depositing a gold layer onto a two-dimensional SiO_2 template fabricated via laser interference lithography and nanoimprint lithography, which is subsequently integrated with PDMS, as illustrated in Figure 3.5. This structure broadens

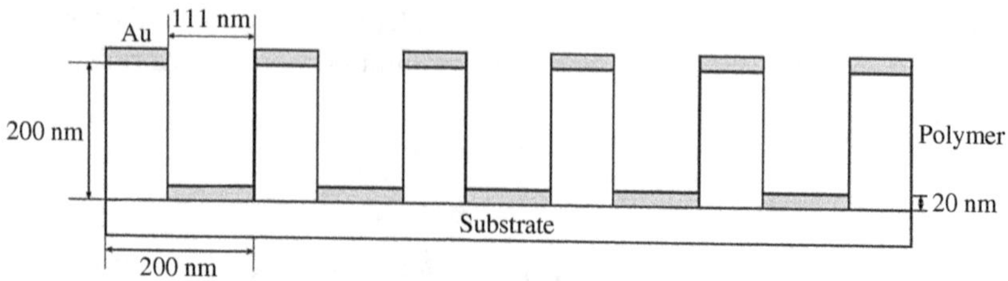

Figure 3.5 Schematic diagram of 2D gold nanostructure.

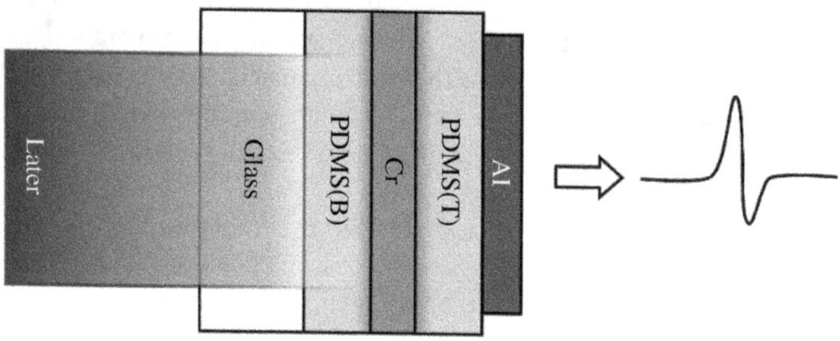

Figure 3.6 Schematic of the sandwich structure of a metal-based LGUS device.

the operational bandwidth of the device, thereby enabling operation above 50 MHz and generating surface acoustic pressures on the order of MPa [15]. In addition, a strategic choice is to pursue a novel device architecture, such as a sandwich structure in which a circular aluminum film array is sputtered onto a PDMS/Cr/PDMS composite, forming a Fabry–Perot (F-P) cavity between the chromium film and the aluminum array elements. This cavity confines light and enhances absorption by the chromium layer, as shown in Figure 3.6. While this design improves LGUS output over a certain range of laser energies, it also increases fabrication complexity [16]. The basic properties of metal-PDMS LGUS devices are shown in Table 3.2.

However, the material's low laser-induced damage threshold and suboptimal energy conversion efficiency pose significant challenges, resulting in a limited achievable acoustic pressure amplitude.

3.1.2 Carbon Nanomaterials-PDMS Devices

With the advancement of research, carbon nanomaterials have attracted increasing attention owing to their outstanding optical and thermal

Table 3.2 Performance summary of representative metal-based LGUS device.

Materials	d(μm)	CF (MHz)	−6 dB BW (%)	$\eta(\times 10^{-2})$	P (MPa)	h (mm)	E (mJ/pulse)	References
Ti	1×10^{-9}	—	—	—	0.9	—	2.35	[17]
Cr	—	—	—	—	3.82	20	20	[17]
Cr sandwiched	1×10^{-9}	—	—	0.42×10^{-3}	1.83	2	2.35	[15]
Au	0.02	~80	~180	—	1.5	0	1×10^{-4}	[14]
Au 2D array	0.03	~40	142.5	—	>1	—	1×10^{-3}	[15]

Materials: light-absorbing materials; BW: bandwidth. CF: center frequency; d: device thickness; E: the energy of pulse laser; h: the distance of hydrophone and device; η: photoacoustic conversion efficiency; P: acoustic pressure.

Table 3.3 Performance summary of a representative carbon-based LGUS device.

Materials	d (μm)	CF (MHz)	−6 dB BW (%)	$\eta(\times 10^{-2})$	P (MPa)	h (mm)	E (mJ/pulse)	References
CSNPs	6	10	210	0.441	+4.8/−1	4.2	3.57	[18]
CB-Lens		14.5	153		+22.5/−8	12.3	18	[3]
CNFs	24.4 ± 1.43		~203	1.56	12.15 ± 1.35	3.65	4.2	[6]
CNT array	18	20.2	152	0.251	+8.8	2	10	[17]
CNTs yarn-Au	5	11.8	179	2.74	+33.6/−10	5	45	[19]
Functionalized CNTs	38	~100	171	—	1.69	0		[20]
rGO	~40	~14.7	~165	—	1.7	1.6	42.3	[21]
Graphite			—		0.48		1.45×10^{-2}	[21]
Graphene	17		—	1.9	2.6	4	60	[22]
Graphene-Au	17			1.2	1.9	4	60	[22]

Materials: light-absorbing materials; BW: bandwidth; CF: center frequency; d: device thickness; E: the energy of pulse laser; h: the distance of hydrophone and device; η: photoacoustic conversion efficiency; P: acoustic pressure.

properties—such as broadband light absorption, high photothermal conversion efficiency, excellent thermal conductivity, low specific heat capacity, and favorable thermal diffusivity. Compared to metal-PDMS composites, LGUS devices based on carbon nanomaterial-PDMS composites exhibit higher photoacoustic conversion efficiency and laser-induced damage threshold. Commonly used carbon nanomaterials include carbon black (CB), carbon nanofibers (CNFs), carbon nanotubes (CNTs), candle-soot nanoparticles (CSNPs), graphite, graphene, and reduced graphene oxide. The basic performance characteristics of representative LGUS devices employing these carbon nanomaterials as light-absorbing layers are summarized in Table 3.3.

3.1.2.1 Carbon Black

As a common and widely accessible black powder, carbon black (CB) was the first carbon nanomaterial adopted as a light-absorbing medium in LGUS devices. It consists mainly of carbon, accompanied by trace amounts of oxygen, hydrogen, and sulfur—a composition that gives rise to a subtle, earthy scent when handled in bulk. CB particles are nearly spherical, presenting as a fine, dark powder that exhibits slight clumping upon agitation. Thanks to its nanoscale architecture, CB offers a high specific surface area, which promotes efficient light absorption and suppresses reflection, yielding a deep, almost matte black appearance that appears to absorb nearly all incident illumination. This material demonstrates strong optical absorption across the visible and near-infrared spectra, capable of capturing most incoming light and thus maintaining uniform darkness under diverse lighting conditions. Moreover, the interconnected nature of CB particles facilitates thermal conductivity (~0.03 $W \cdot m^{-1} \cdot K^{-1}$), permitting heat to transfer relatively efficiently through the material.

In practice, CB is commonly integrated with a thermoelastic polymer such as PDMS to fabricate functional LGUS transducers. The composite is typically prepared via spin-coating followed by thermal curing, producing a smooth, flexible film. The concentration of CB within the PDMS matrix critically influences both rheological and optical properties: overly dilute solutions lead to insufficient laser absorption and consequently weak ultrasound generation, whereas increasing CB content raises viscosity, often resulting in nonuniform coatings characterized by visible streaks or thickness variations. At concentrations exceeding about 61.28%, the mixture may fail to cure properly, leaving the film tacky and mechanically unstable [6]. Compared with conventional metal-film-based designs,

CB-PDMS composite devices deliver markedly stronger LGUS signals, producing clearer and more intense acoustic pulses. Advances in the processing of CB-PDMS composites have allowed these devices to generate peak sound pressures on the order of MPa, paving the way for their practical application.

Nevertheless, two primary material-level challenges stem from the particulate morphology of CB: stability and heterogeneity. The physical form and structure of CB affect its stability under operational conditions. Notably, high-temperature or intense optical irradiation can cause structural degradation (e.g. particle aggregation/fragmentation), consequently impairing the reproducibility of the photoacoustic response and leading to inconsistent signals. Simultaneously, the inherent nonuniformity in particle size and distribution often results in localized clustering. This heterogeneity complicates fabrication by creating uneven light absorption and acoustic generation, which in turn threatens long-term performance stability due to potential particle rearrangement.

3.1.2.2 Carbon Nanofibers

Carbon nanofibers (CNFs) are cylindrical or conical nanomaterials composed of stacked graphene cones, with diameters ranging from a few nanometers to several hundred nanometers and lengths on the micrometer scale. They are typically produced by electrospinning of organic polymers followed by thermal treatment in an inert atmosphere [23]. CNFs not only retain the intrinsic properties of conventional carbon fibers but also exhibit a high surface area-to-volume ratio, along with abundant surface micropores and nanoscale roughness. This structure enables efficient light absorption. Moreover, strong covalent bonds and tightly stacked graphene layers facilitate rapid heat conduction within CNFs. Benefiting from strong light absorption and an extremely high thermal conductivity ($\approx$1950 W m^{-1} K^{-1}), CNFs demonstrate considerable potential for efficient light-to-heat energy conversion.

CNFs provide a cost-effective and robust platform for LGUS devices. The device fabrication follows a sequential workflow. First, precursor fibers generated from a polyacrylonitrile (PAN) solution were electrospun and thermally treated (stabilized at 250 °C and carbonized at 900 °C) to form a 25-μm thick CNFs film on a glass substrate. Following the process transitioned to the polymer coating stage. Subsequently, a PDMS layer was spin-coated onto the CNFs film to encapsulate the structure and enhance its acoustic performance. The resulting CNFs-PDMS-based LGUS device delivers a peak sound pressure of 12.15 ± 1.35 MPa with a −6 dB bandwidth of 7.63 MHz. This corresponds to a 7.6-fold (17.62 dB) increase in maximum acoustic pressure compared to devices based on CB-PDMS films. Owing to its high energy conversion efficiency and excellent performance, this device holds significant promise for applications in ultrasound imaging and therapy, paving the way toward cost-effective and high-efficiency LGUS designs [6].

3.1.2.3 Candle-Soot Nanoparticles

The soot that drifted when humans first lit candles may have inadvertently given birth to the earliest candle-soot nanoparticles (CSNPs) studies have confirmed that CSNPs are small in size and highly adsorbent. At the beginning of the 21st century, materials scientists discovered that CSNPs excel in physical applications as a treasure nanomaterial that is one of the raw materials for superhydrophobic anti-icing coatings. In 2015, CSNPs were first applied as a light-absorbing material in an LGUS device [18].

CSNPs are spherical carbon nanomaterials produced during candle burning, which can be obtained by placing a glass plate above the candle flame at a very low cost and are easy to prepare. They have a low specific heat capacity (~750 J K^{-1} Kg^{-1}). Their nanoscale size and layered and dendritic structure of CSNPs endow them with very high surface-to-volume ratios, which promotes heat transfer with thermal expanders and contributes to efficient thermal diffusion. Due to their low interfacial thermal resistance property, heat is conducted efficiently between the nanoparticles and the thermally expandable material, which can form a significant temperature gradient, and thus the CSNPs composites have a high photoacoustic conversion efficiency, with a light absorption coefficient of up to about 96% (96.24% ± 0.31%) [24].

The fabrication of specific CSNPs-based LGUS devices follows a straightforward procedure, as illustrated in Figure 3.7. First, CSNPs are collected above the candle flame, and when the flame height reaches about 3 cm, CSNPs are collected on a glass slide about 2 cm above the wick, and the thickness of CSNPs layer on the glass slide is determined by the deposition time. Then, PDMS is spin-coated directly onto the CSNPs-laden slides. The composite structure is then thermally cured at 85 °C for 1 hour [25]. In terms of performance, the device utilizing CSNPs as the light-absorbing material achieves a peak acoustic pressure of 4.8 MPa. It also demonstrates excellent frequency characteristics, with a −6 dB bandwidth of 21 MHz and a center frequency of 10 MHz [18]. In conclusion, this fabrication

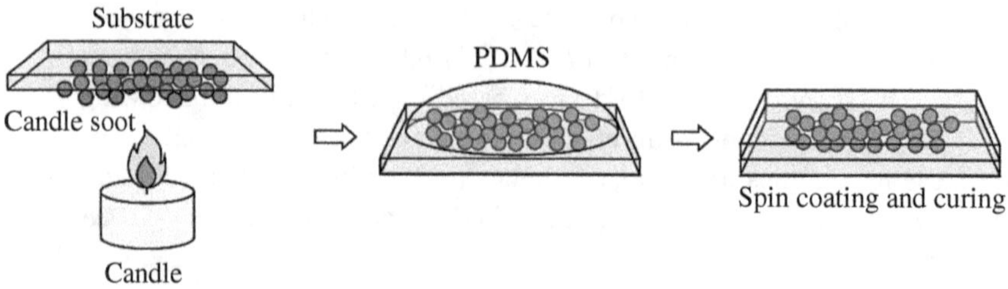

Figure 3.7 Schematic diagram of the process of collecting CSNPs on a glass plate.

strategy offers a simple, cost-effective, and reliable pathway for developing LGUS devices with significant potential for biomedical applications of ultrasound.

3.1.2.4 Carbon Nanotubes

CNTs are one-dimensional quantum materials with a unique structure, featuring radial dimensions on the order of nanometers and axial dimensions on the order of micrometers. Structurally, they can be visualized as graphene sheets rolled into seamless cylinders. Based on the number of rolled layers, CNTs are classified into single-walled CNTs (SWCNTs), which consist of a single layer of carbon atoms arranged in a hexagonal honeycomb lattice, as shown in Figure 3.8 (a), and multi-walled CNTs (MWCNTs), which comprise multiple coaxial tubes with a fixed interlayer spacing, as shown in Figure 3.8(b).

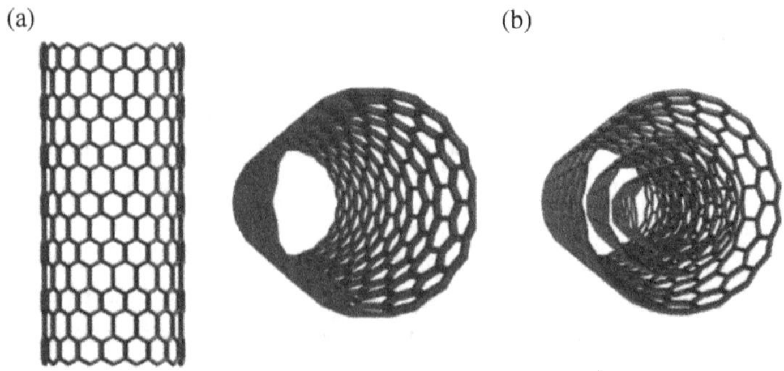

Figure 3.8 Schematic representation of single-walled carbon nanotubes (a) and multi-walled carbon nanotubes (b).

Both SWCNTs and MWCNTs exhibit excellent light absorption, primarily attributed to their unique electronic structure and sp^2 hybridized carbon–carbon bonding. Although their diameters are much smaller than the wavelengths of visible and infrared light, CNTs interact strongly with electromagnetic radiation, and their absorption properties can be tuned by modifying their chirality, diameter, and surface functionalization.

In terms of thermal properties, SWCNTs demonstrate an exceptionally high thermal conductivity of approximately 3500 W m^{-1} K^{-1} [9]. This combination of optical and thermal performance makes CNTs ideal candidates for LGUS applications, where generated sound pressures can reach the MPa level.

For device fabrication, a typical approach involves the direct growth of MWCNTs on fused silica substrates via chemical vapor deposition (CVD), followed by coating with a thermally expandable PDMS layer. As shown in Figure 3.9, this results in a robust three-layer structure. Such CNT-PDMS composite devices exhibit superior stability; for instance, they can withstand laser fluences up to 477 mJ cm^{-2}, demonstrating a significantly higher damage threshold compared to other composites [26].

Due to the unique molecular structure of CNTs, their dimensions significantly influence their thermal properties. Specifically, the thermal conductivity of CNTs typically increases with length, eventually saturating at a stable value, whereas it tends to be inversely proportional to diameter. Furthermore, the dimensions of CNTs modulate their specific heat capacity and thermal diffusivity. These factors collectively determine the thermal transfer efficiency between the light-absorbing layer and the thermal expansion layer, thereby governing the output acoustic pressure of the LGUS device [27]. Consequently, to optimize heat transfer and enhance energy conversion efficiency, CNTs with longer lengths (20–30 μm) and smaller diameters (~5 nm) are preferred. Utilizing CNTs with these optimized dimensions, an LGUS device was fabricated that produced high-intensity ultrasound with a peak pressure of approximately 24.4 MPa. This device

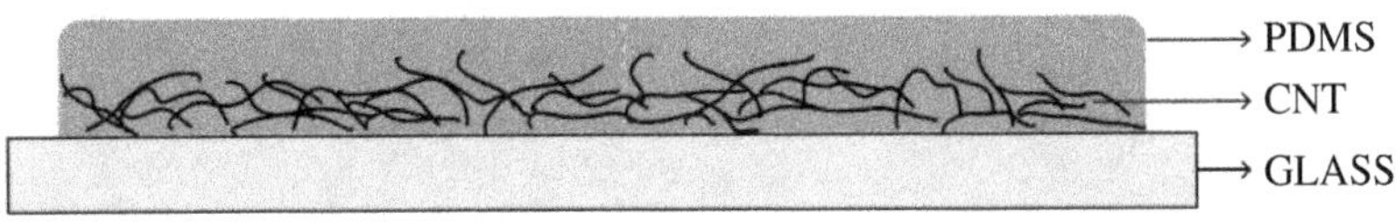

Figure 3.9 Schematic of a three-layer structure of an LGUS device based on CNTs.

exhibited a center frequency of 7.8 MHz, a −6 dB bandwidth of 13.5 MHz, and a photoacoustic conversion efficiency of about 9.59×10^{-3}.

Biomedical applications such as thrombolysis, cavitation, and ablation require high acoustic pressures, necessitating the development of LGUS devices with enhanced output intensity. To achieve this, CNTs can be composited with noble metals to augment their light absorption properties. Specifically, the localized surface plasmon resonance (LSPR) effect of gold nanoparticles (AuNPs) significantly boosts optical absorption, thereby improving the photothermal conversion efficiency of the device.

A representative high-performance device is fabricated by depositing a layer of gold nanoparticles onto CNT yarns, followed by spin-coating with PDMS, as illustrated in Figure 3.10. This specific PDMS/Au-CNT yarn composite structure utilizes the synergistic effect of the materials to achieve sound pressures as high as 33.6 MPa (at a fluence of $45\,\mathrm{mJ\,cm^{-2}}$) and an energy conversion efficiency of 2.74×10^{-2} [19].

The frequency of ultrasound is a key parameter for biomedical applications, necessitating precise frequency modulation capabilities in LGUS devices. This can be achieved by manipulating the geometric dimensions of the CNTs composite layers.

In planar devices, the operating frequency is governed by the thickness of the absorbing layer. For instance, vertically aligned CNTs arrays can be grown via chemical vapor deposition (CVD) and subsequently thinned to precise dimensions using inductively coupled plasma (ICP) etching. In these devices, which leverage the anisotropic thermal conductivity of aligned CNTs, the operating frequency is inversely proportional to the thickness of the CNTs composite [27].

Moreover, frequency modulation can be achieved by altering the diameter of the photoacoustic source. As shown in Figure 3.11, polyethylene microspheres coated with carbon nanotubes (MNCNTs) can be immobilized on

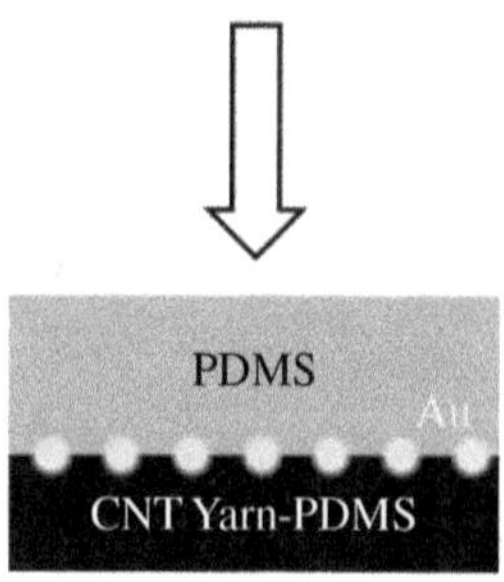

Figure 3.10 PDMS/Au-CNT Yarn-PDMS structure: the light-absorbing material is a multilayer CNT yarn with gold particles deposited.

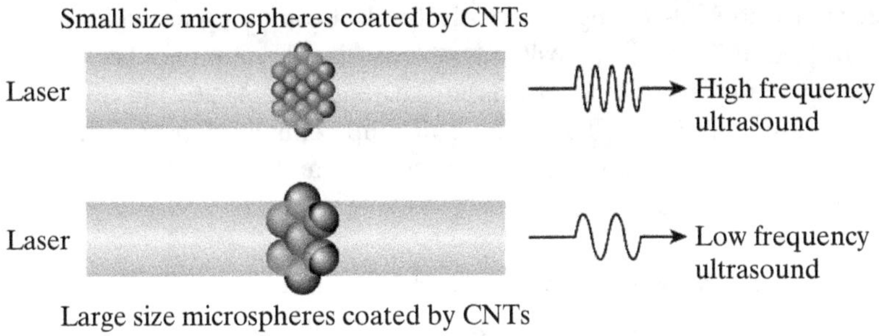

Figure 3.11 MNCNTs were coated on the surface of polyethylene microspheres, which were immobilized on PDMS or glass, and the frequency was modulated by modulating the diameter of the microspheres.

a PDMS or glass substrate. In this configuration, the center frequency is tunable by adjusting the diameter of the microspheres: smaller microspheres generate higher-frequency ultrasound due to their faster thermal expansion and contraction cycles [28].

Due to strong intermolecular van der Waals forces, CNTs exhibit a pronounced tendency to aggregate and possess poor solubility in common organic solvents. This makes it difficult to achieve a uniform distribution within the PDMS matrix, posing a significant challenge for the fabrication of consistent LGUS devices. To address this, two primary strategies are employed to enhance the dispersion and solubility of CNTs.

First, surface functionalization with oleylamine can significantly improve the solubility of CNTs in xylene while preventing aggregation. The subsequent addition of acetone induces the formation of an organogel, ensuring the homogeneity of the CNTs-PDMS-based photoacoustic composites [29].

Second, CNT surfaces can be modified via acyl chlorination to introduce reactive acyl chloride functional groups (-COCl-MWCNTs) [20]. This modification not only enhances the binding affinity between the CNTs and the polymer matrix (e.g. PDMS) but also improves their dispersion in polystyrene/tetrahydrofuran solutions. Critically, these functionalization techniques augment the thermal diffusivity and light absorption properties of the composites. Consequently, the resulting CNTs films are capable of generating intense LGUS signals under laser irradiation, while exhibiting improved fatigue resistance and mechanical stability.

Beyond planar devices, fabricating miniaturized devices requires specialized coating techniques. For instance, to achieve a uniform CNTs composite coating on the end-face of optical fibers, dip-coating serves as an effective alternative to spin-coating. Alternatively, electrospinning technology can be utilized to deposit composite nanofibers directly onto the fiber tip. By refining these preparation methods and coating technologies, the efficiency of composite fabrication has been significantly enhanced, paving the way for the development of low-cost, miniaturized, and high-efficiency LGUS devices for clinical applications.

3.1.2.5 Graphite, Graphene, and Reduced Graphene Oxide

Graphite, graphene, and reduced graphene oxide (rGO) form a family of carbon materials whose structures are closely related yet whose physicochemical properties differ markedly. As a naturally occurring layered allotrope of carbon, graphite serves as the bulk precursor for graphene and graphene oxide. By chemical, mechanical, or thermal exfoliation, these derivatives—graphene and rGO—can be effectively isolated from graphite.

Together, graphite and its derivatives constitute an important class of light-absorbing materials, whose performance is strongly coupled to their distinctive layered atomic architectures. Owing to variations in carbon-layer stacking and interlayer interactions, these materials exhibit tunable thermal characteristics, underscoring the critical role of structural factors in governing their light absorption and heat-related behavior.

As shown in Figure 3.12, graphite is characterized by a layered crystalline structure composed of carbon atoms. Within each layer, carbon atoms are

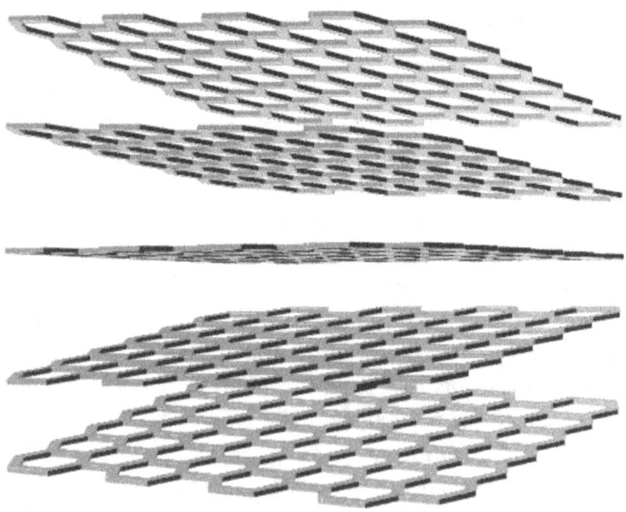

Figure 3.12 Schematic diagram of a graphite structure.

arranged in a planar hexagonal lattice, bonded covalently to three neighboring atoms via sp^2 hybridization. These individual layers act as infinitely extended π-conjugated systems capable of absorbing light across a broad spectrum. The layers are stacked and held together by weak van der Waals forces. This specific electronic structure enables graphite to function as an effective light absorber; as light penetrates the material, it is continuously absorbed and re-absorbed by the successive layers. Furthermore, graphite exhibits high thermal conductivity, attributed to the strong covalent bonds between carbon atoms and high in-plane electron mobility, which facilitates rapid heat transfer. Owing to its high optical absorption efficiency and thermal conductivity, graphite can effectively absorb diffuse laser light and convert it into thermal energy. Consequently, LGUS devices utilizing graphite as the light-absorbing material can generate maximum sound pressures reaching the MPa level, sufficient for applications such as ultrasonic nerve stimulation [21].

Graphene can be conceptualized as a single layer of graphite. As shown in Figure 3.13, it is a two-dimensional material consisting of a single layer of carbon atoms arranged in a hexagonal lattice with sp^2 hybridization. In 2010, the Nobel Prize in Physics was awarded to A. Geim and K. Novoselov from the University of Manchester for their groundbreaking experiments in 2004, where they successfully isolated single-layer graphene using mechanical exfoliation with adhesive tape.

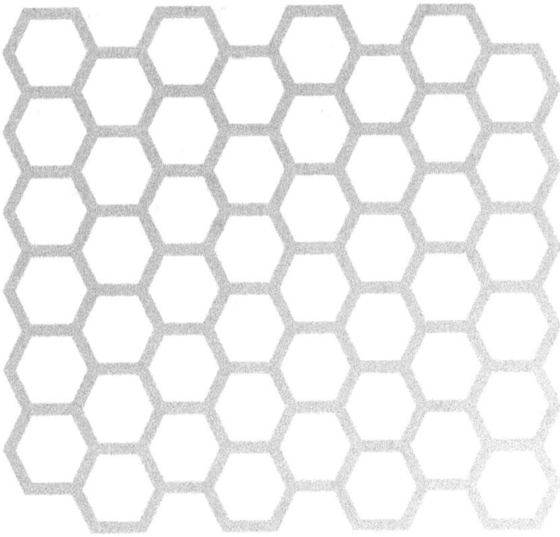

Figure 3.13 Schematic representation of a graphene structure.

In this structure, electrons are free to move within the plane, endowing graphene with exceptionally high electrical conductivity and optical responsiveness. Due to its unique electronic band structure, graphene exhibits broadband light absorption, particularly efficient in the visible and infrared regions. While bulk graphite's absorption is limited by scattering effects between layers, monolayer graphene demonstrates remarkable opacity relative to its atomic thickness. Moreover, the thermal conductivity of graphene is extremely high due to the strong covalent bonding and high electron mobility in the 2D plane, allowing for rapid heat conduction. Its monolayer nature also helps in minimizing interfacial thermal resistance when properly integrated.

As an excellent photoacoustic material, 2D graphene composites combined with thermally expandable polymers can be used to fabricate LGUS devices achieving pressures of approximately 2.6 MPa with a −6 dB bandwidth of about 40 MHz. To broaden the bandwidth for diverse biomedical applications, graphene can be modified in situ with gold nanoparticles (AuNPs). This modification increases the thermal interface area and allows for precise control over the film thickness. Devices fabricated based on these AuNP-modified films can achieve a bandwidth of approximately 50 MHz under similar conditions [22].

Reduced graphene oxide (rGO) is a derivative of graphene that retains the two-dimensional layered structure, where carbon atoms are arranged in a hexagonal lattice with sp^2 hybridization. It is produced by the reduction of graphene oxide, a process in which oxygen-containing functional groups (e.g. carboxyl, hydroxyl, and epoxy groups) are partially removed, resulting in a restoration of the ordered carbon arrangement. rGO is particularly suitable for LGUS devices due to its high optical absorption (spanning from UV to infrared), high thermal conductivity (~300 $W\,m^{-1}\,K^{-1}$), and nanoscale dimensions. To further enhance performance, composite films have been developed using carbon black doping to augment the optical absorption of rGO. Such devices have demonstrated acoustic pressures of up to 1.7 MPa, with a −6 dB bandwidth of approximately 24.3 MHz and a center frequency of 14.7 MHz [12].

3.1.3 Other Composite Material Devices

Beyond metals and carbon nanomaterials, a variety of other materials exhibit excellent optical and thermal properties suitable for LGUS applications. Table 3.4 summarizes the fundamental properties and performance metrics of LGUS devices fabricated using these alternative materials.

Table 3.4 Performance summary of representative other material LGUS device.

Materials	d (μm)	CF (MHz)	−6 dB BW (%)	η ($\times 10^{-2}$)	P (MPa)	h(mm)	E (mJ/pulse)	References
Ink (Lumocolor)		~14	~123		0.266	1.5	60	[30]
CIS		~10	~180		3.7	1.5	48.2	[31]
Epolight 9837					1.7	1.5	15.9	[32]
$MAPbI_3$	0.34	29.2	140	2.97	+15/−10	2	3	[11]
MXene	3			1.25	15.7		42	[33]

Materials: light-absorbing materials; d: device thickness; CF: center frequency; BW: bandwidth; η: photoacoustic conversion efficiency; P: acoustic pressure; h: distance between hydrophone and device; E: pulse laser energy.

3.1.3.1 Ink

Permanent marker ink (e.g. Lumocolor, Staedtler, UK) is a commercially available, low-cost material exhibiting wavelength-selective absorption. This ink can be processed into a light-absorbing composite suitable for LGUS fabrication. The typical preparation process involves: (i) exposing the ink to air to evaporate the original solvent, followed by redissolving the dried ink pigment in xylene; (ii) mixing this xylene-based ink solution with PDMS to form a homogeneous composite; and (iii) immersing a multimode fiber into the mixture to coat its distal tip (after stripping the buffer layer). This approach enables the composite to absorb specific wavelengths for ultrasound generation while allowing other wavelengths to pass through for photoacoustic excitation. Devices fabricated using this method have demonstrated a −6 dB bandwidth of approximately 17.3 MHz, a center frequency of 14 MHz, and sound pressures reaching the MPa level [30].

3.1.3.2 Quantum Dot Material $CuInS_2$ (CIS)

The development of quantum dot (QD) technology traces back to the 1970s, originating from research into semiconductor–liquid interfaces for energy applications. In 1981, physicists successfully synthesized water-soluble cadmium sulfide colloids. Subsequently, Dr. Brus's team discovered a direct correlation between the particle size of these nanoparticles and their exciton energies, indicated by color changes. This led to the proposal of the "colloidal quantum dot" concept, elucidating the quantum confinement effect. By the late 1990s, these nanomaterials transitioned from energy research to the life sciences, as Alivisatos and Nie's teams successfully developed water-soluble, bioconjugated quantum dots capable of cellular imaging, sparking a surge in nanotechnology research.

Quantum dots (QDs) possess favorable properties for multimodal imaging, including high photostability and tunable absorption profiles. Their narrow absorption spectra can be precisely adjusted by manipulating particle size and composition to maximize absorbance at the incident laser wavelength. Furthermore, QDs can be functionalized for seamless integration into polymer matrices like PDMS.

Among QD materials, copper indium sulfide (CIS) is particularly suitable for biomedical applications due to its lower toxicity compared to heavy-metal-containing QDs (e.g. lead or cadmium). Through ligand engineering, the photoluminescence quantum yield of CIS can be suppressed, thereby enhancing the non-radiative decay pathway. This ensures that a greater proportion of absorbed light energy is converted into heat via the photoacoustic effect. CIS-PDMS composite coatings can generate ultrasonic pressures exceeding 3 MPa with bandwidths greater than 15 MHz, which is critical for high-resolution imaging. Owing to their wavelength selectivity, CIS composites support dual-modality imaging—photoacoustic (PA) imaging and optoacoustic ultrasound (OpUS) imaging—making them highly promising for clinical diagnostics [31].

3.1.3.3 Photostable Dyes

Near-infrared absorbing dyes, such as Epolight 9837 (Epolin, USA), are distinguished by their excellent photostability and high absorption coefficient at 1064 nm, allowing for efficient conversion of laser energy into acoustic energy. Experimental characterization of coatings incorporating this dye indicates high optical absorption. Single fiber-optic LGUS devices fabricated with Epolight 9837 have produced sound pressures up to 2.70 MPa with a −6 dB bandwidth of approximately 28.5 MHz, demonstrating their potential for minimally invasive imaging applications [32].

3.1.3.4 Perovskite ($MAPbI_3$)

The history of perovskite materials began in 1839 when the German mineralogist Gustav Rose discovered calcium titanate ($CaTiO_3$) in the Ural Mountains and named it "Perovskite" after the Russian mineralogist Lev Perovsky. While early research focused on crystallography, application-oriented studies expanded significantly in the mid-20th century. A major breakthrough occurred in 2009 when Miyasaka's team utilized methylammonium lead iodide ($CH_3NH_3PbI_3$) as a photosensitizer to fabricate the first perovskite solar cell (3.8% efficiency). This initiated a paradigm shift in advanced photovoltaic research. Today, the perovskite family—defined by the ABX_3 structure (where A is an organic cation, B is Pb^{2+}/Sn^{2+}, and X is halogen anion)—has expanded beyond photovoltaics into LEDs, radiation detection, and quantum computing, owing to its exceptional light absorption coefficient (on the order of 10^5 cm^{-1}), solution processability, and tunable bandgap (1.5–2.3 eV).

Recently, perovskite-based materials have garnered attention for LGUS applications due to their favorable thermal properties, including low specific heat capacity (296–326 J K^{-1} Kg^{-1}), low thermal conductivity, and small thermal diffusivity (0.145 mm^2 s^{-1}), combined with high optical absorption. LGUS devices utilizing $MAPbI_3$ as the light-absorbing layer have demonstrated high performance, achieving a peak sound pressure of approximately 2.72 MPa and a high energy conversion efficiency of

2.97×10^{-2}. These devices exhibit a center frequency of 29.2 MHz and a −6 dB bandwidth of 40.8 MHz. The fabrication of such thin-film LGUS devices on optical fibers has enabled high-resolution imaging capabilities, sufficient to resolve fine structures such as the fish eye [11].

3.1.3.5 2D Transition Metal Carbide/Nitride (MXene)

MXenes, a class of two-dimensional inorganic compounds comprising transition metal carbides, nitrides, or carbon nitrides, were first reported in 2011. These materials consist of layers only a few atoms thick and exhibit metallic conductivity due to surface termination groups (e.g. hydroxyl or oxygen). MXenes possess strong absorption in the visible and near-infrared regions, making them highly effective for optoelectronic applications. Notably, $Ti_3C_2T_x$, a common MXene, demonstrates excellent photothermal conversion properties, with an internal conversion efficiency approaching 100%. This efficiency is largely attributed to the localized surface plasmon resonance (LSPR) effect. Given these advantages, MXenes are being explored for localized cancer therapy, photothermal catalysis, solar desalination, and wearable electronics. In the context of LGUS, a device based on a glass/MXene ($Ti_3C_2T_x$)/PDMS sandwich structure has been shown to efficiently convert laser pulses into ultrasonic waves, generating a high acoustic pressure of 15.7 MPa with an efficiency of 1.25×10^{-2} [33].

3.2 Thermal Expansion Material

Following the efficient absorption of optical energy, the subsequent and equally critical stage in LGUS involves the transduction of thermal energy into mechanical waves. This process is mediated by the thermal expansion material, which functions as the acoustic engine of the transmitter. As detailed in the theoretical framework of Chapter 2, the efficiency of this thermoelastic generation is fundamentally driven by the material's ability to undergo rapid volumetric changes in response to temperature fluctuations. Consequently, the volume thermal expansion coefficient (CTE) serves as the primary figure of merit for selecting suitable materials; a higher CTE translates directly to a more significant volume expansion for a given thermal input, thereby generating higher amplitude acoustic pressures. Beyond high expansion capability, the material must possess optical transparency at the excitation wavelength to prevent premature energy loss and an acoustic impedance matched to the propagation medium to facilitate efficient wave coupling.

The architectural design and constituent materials of LGUS devices are critical determinants of photoacoustic conversion efficiency. Historically, LGUS technology has evolved from simple inorganic films to sophisticated inorganic–organic composite structures. Initial investigations focused on single-layer metallic films, which possess both light absorption and thermal expansion properties, enabling them to complete the photoacoustic conversion process independently. However, metal surfaces exhibit high optical reflectivity and relatively weak volumetric thermal expansion, resulting in suboptimal photoacoustic conversion efficiency [34].

To surmount these thermodynamic limitations, the field shifted toward inorganic–organic composite structures. These composites consist of a distinct light-absorbing layer (typically inorganic nanomaterials) and a thermal expansion layer (typically organic polymers). Upon excitation, the inorganic material converts light into heat, which is rapidly transferred to the adjacent organic material. The organic layer, leveraging its superior coefficient of thermal expansion, converts this thermal energy into acoustic energy, thereby generating high-amplitude ultrasonic signals. In practical applications, the rapid mechanical response of the expansion material directly influences the bandwidth and pressure of the LGUS signal.

Therefore, the characterization and optimization of thermal expansion materials tailored to specific application requirements are pivotal for advancing LGUS technology. Currently, polymeric materials dominate this field due to their superior thermoelastic efficiency compared to metals or inorganic solids. The polymers currently applied in the LGUS field include PDMS, Polymethyl Methacrylate (PMMA), and Ecoflex silicone.

3.2.1 Polymethylmethacrylate (PMMA)

Polymethylmethacrylate (PMMA), a high-molecular-weight polymer also known as acrylic or plexiglass, has a chemical formula of $(C_5O_2H_8)_n$. It is characterized by exceptional transparency (light transmittance up to 92%, surpassing glass), cost-effectiveness, and ease of mechanical processing, with a thermal expansion coefficient of $90 \times 10^{-6/}$°C [35]. The polymerization of PMMA involves the catalytic initiation of double bonds in methyl methacrylate monomers, forming a macromolecular chain structure. Its synthesis is relatively straightforward, allowing for the regulation of molecular weight and structure to meet diverse application needs.

The history of PMMA dates back to 1901, when German chemist Otto Röhm first synthesized it experimentally. However, large-scale production was initially impractical. In 1928, British chemist Rowland Hill advanced the research, highlighting that PMMA possessed glass-like transparency with significantly lower density. In 1933, the German firm Rohm and Haas commercialized PMMA under the trademark "Plexiglas," a milestone that paved the way for its eventual adoption in LGUS technology.

In contemporary applications, an all-optical intravascular ultrasound (AO-IVUS) imaging system has been developed using a picosecond laser to excite ultrasound in Carbon-PMMA composites, coupled with a π-phase shift fiber Bragg gratings (π-FBG) for detection. This system achieves an ultrawide bandwidth of 147% and a high axial resolution of 18.6 μm, outperforming many traditional technologies. The device utilizes a 5 μm-thick MWCNT-PMMA composite film shaped into a concave structure with a 3-mm radius of curvature to focus the ultrasonic beams. Phantom studies demonstrated an axial resolution of 18.6 μm, a lateral resolution of 124 μm, and an imaging depth of 7 mm. Furthermore, rotational pullback imaging in rabbit iliac arteries, porcine coronary arteries, and stented rabbit arteries confirmed its capability to depict fine vascular details [36]. Despite these successes, PMMA's relatively low thermal expansion coefficient limits its conversion efficiency.

3.2.2 Polydimethylsiloxane (PDMS)

PDMS, a high-molecular polymer with the chemical formula $(C_2H_6OSi)_n$, features a large thermal expansion coefficient ($\beta = 930 \times 10^{-6}$ K^{-1}). This property enables highly efficient absorption of thermal energy and subsequent generation of ultrasonic signals during photoacoustic conversion. Additionally, PDMS combines transparency with excellent plasticity, allowing it to be molded into complex geometries and recover from deformation, thus satisfying diverse design requirements.

The development of organosilicon chemistry began in 1924 when Kipping synthesized the first polysilane via sodium condensation of diphenyldichlorosilane, though the product was insoluble and difficult to process. Progress stalled until 1949 when Burkhard synthesized polydimethylsilane, yet processing issues persisted. A breakthrough occurred in 1975 when Seishi Yajima converted polydimethylsilane into soluble polycarbosilane, reigniting interest in the field. By the 1980s, the synthesis of soluble polysilanes had advanced rapidly, leading to various

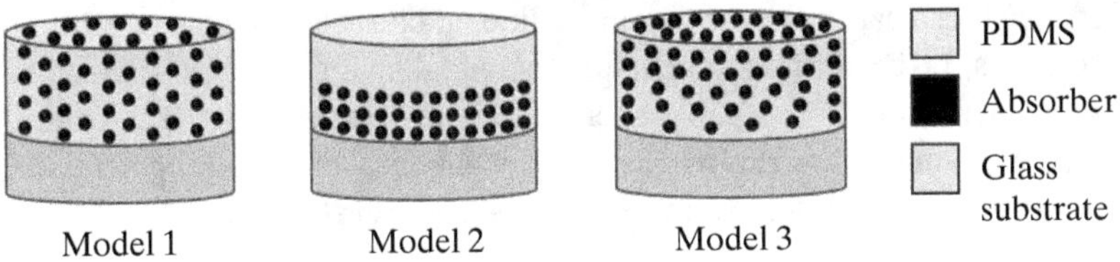

Figure 3.14 Composite methods of PDMS.

copolymers and block copolymers with tailored properties. Today, PDMS is widely regarded as a benchmark material in LGUS due to its superior thermoelastic properties.

3.2.2.1 Composite Methods of PDMS

To fabricate LGUS devices, optical absorbers (e.g. CNTs, metal nanoparticles) must be integrated with PDMS. Compared to absorber-only coatings, absorber-PDMS composites yield higher acoustic pressures and higher damage thresholds. Fabrication strategies are generally categorized into three modes based on the spatial distribution of the absorbers, as shown in Figure 3.14:

Mode 1: Mixing light absorber nanoparticles directly with the PDMS precursor. While simple and versatile, this method can suffer from aggregation issues and inefficient utilization of expensive nanomaterials buried deep within the matrix where light may not reach.

Mode 2: Depositing a thin absorption layer on a substrate followed by a PDMS overlay. This allows precise control over the absorption layer thickness. However, techniques like CVD may be incompatible with temperature-sensitive substrates or complex geometries (e.g. optical fiber tips).

Mode 3: Incorporating optical absorbers into a pre-cured or partially cured PDMS film. This can be achieved via diffusion from solution or ion implantation. Top-down approaches allow the PDMS surface to be optimized before absorber integration, potentially minimizing material waste [37].

3.2.2.2 Optimization of PDMS—Self-healing PDMS

While LGUS devices are valuable in biomedical applications, they are susceptible to mechanical damage (e.g. scratches) during operation. Self-healing materials offer a solution to enhance reliability and extend service life, a capability lacking in traditional piezoelectric devices. Unlike ceramics, LGUS devices are polymer-based, allowing for the introduction of reversible or exchangeable chemical bonds. Upon mechanical failure, these dynamic bonds can reconstruct the chemical cross-linking network, restoring material integrity and ensuring long-term stability.

3.2.2.3 Preparation Process and Structural Characteristics of Self-Healing PDMS

A prominent example of self-healing PDMS is Fe-Hpdca-PDMS. The preparation involves a condensation reaction between bis(3-aminopropyl)-terminated PDMS (H_2N–PDMS–NH_2, Mn = 5000–7000) and 2,6-pyridinedicarbonyl dichloride to form a PDMS oligomer containing 2,6-pyridinedicarboxamide (H_2pdca) ligands. This oligomer is then cross-linked with iron(III) chloride ($FeCl_3$) at a 1:2 molar ratio of Fe(III) to ligands, yielding a dark red solid [7].

3.2.2.4 Realization Mechanism of PDMS Self-Healing

The self-healing mechanism relies on the dynamic nature of metal–ligand interactions. In Fe-Hpdca-PDMS, intramolecular interactions cause chain folding, while intermolecular interactions create a 3D network. As shown in Figure 3.15, when the material is damaged (e.g. cut), the coordination bonds in the $[Fe(Hpdca)_2]^+$ complex break. The polymer chains unfold and slide, but the iron ions remain tethered to the ligands via the strong pyridine interactions, forming an intermediate $Fe(Hpdca)]^{2+}$ state.

Upon removal of stress or bringing the cut surfaces together, the polymer chains refold, and the ligands re-chelate with the iron centers, reforming the original $[Fe(Hpdca)_2]^+$ complex and healing the material [38]. The material's low glass transition temperature ($T_8 < -90\,°C$) ensures high chain mobility even at low temperatures. Experimental results show that after 48 hours at room temperature, cuts are virtually repaired with 90% efficiency. Even at $-20\,°C$, repair efficiency reaches 68% after 72 hours. This technology offers a promising avenue for creating durable LGUS devices.

3.2.3 Ecoflex

While PDMS has long served as the standard polymeric expansion layer in LGUS devices due to its high coefficient of thermal expansion, emerging ultrasoft elastomers such as Ecoflex have demonstrated superior capabilities for specific high-efficiency applications. Distinguished by its exceptionally

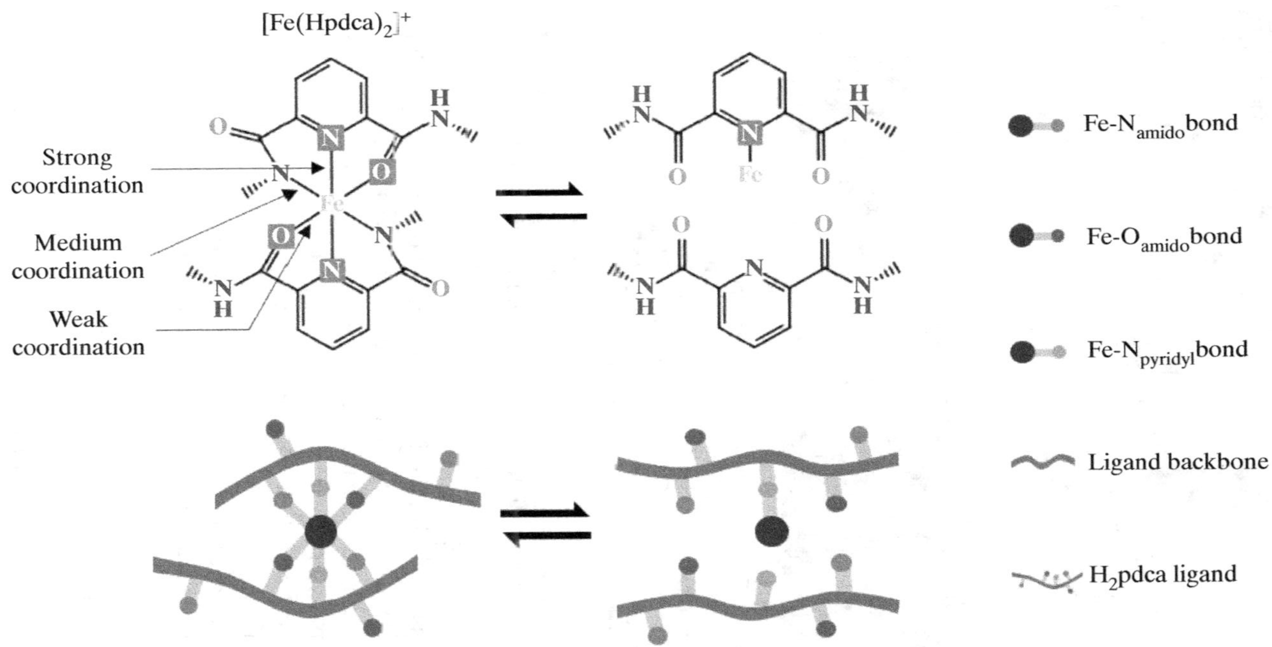

Figure 3.15 Structural and schematic diagram of [Fe(Hpdca)$_2$]$^+$.

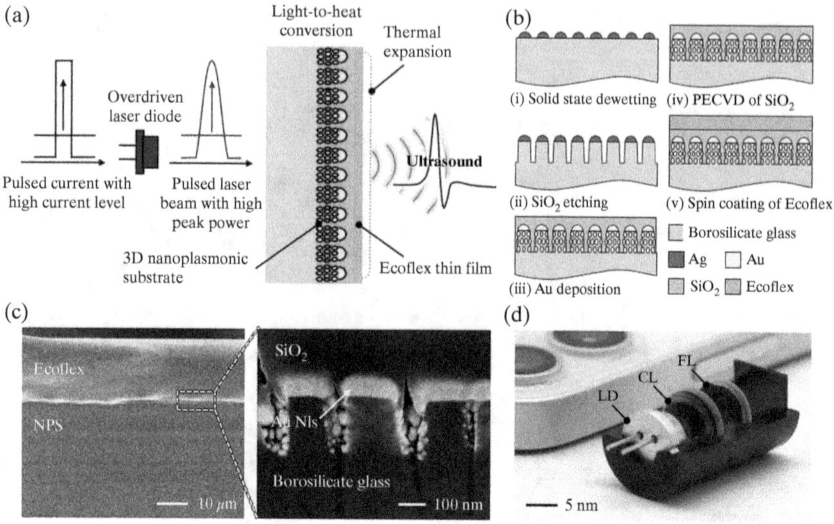

Figure 3.16 PEAT using focused pulsed laser beam. (a) A schematic illustration of PEAT including Ecoflex-coated NPS. (b) Nanofabrication procedures of Ecoflex-coated NPS using solid state dewetting of Ag thin film, reactive ion etching, Au thin film evaporation, PECVD of SiO_2, and spin-coating of the thin Ecoflex film. (c) Cross-sectional SEM images of the Ecoflex-coated NPS. (d) Fully packaged PEAT module consisting of an LD, a CL, an FL, and an Ecoflex-coated NPS.

low modulus and a CTE surpassing that of PDMS, Ecoflex has emerged as a promising candidate for next-generation LGUS devices. As shown in Figure 3.16, Na et al. reported a compact plasmon-enhanced optoacoustic transducer (PEAT) that integrates an overdriven pulsed laser diode with a 3D nanoplasmonic substrate coated with a thin Ecoflex film. By optimizing the Ecoflex thickness to 22 μm to facilitate rapid heat transfer, the device achieved an eightfold enhancement in optoacoustic amplitude relative to a bare substrate, along with a broad fractional -6 dB bandwidth exceeding 160%. This configuration not only enabled precise control of acoustic pressure via laser modulation but also successfully demonstrated acoustic atomization in microfluidic chips, highlighting the material's efficacy for high-efficiency broadband ultrasound generation [39].

3.3 Summary

This chapter has systematically explored the foundational material strategies for LGUS devices, specifically focusing on the optimization of light-absorbing materials and thermal expansion materials to enhance photoacoustic conversion efficiency. Regarding light-absorbing materials, we traced the technological evolution from traditional metallic films (e.g. Au and Cr) to advanced carbon nanomaterials. While metals offer ease of fabrication, carbon-based variants such as CNTs, CSNPs, and graphene have emerged as superior alternatives due to their exceptional optical absorption and rapid thermal diffusion capabilities. Furthermore, the chapter highlighted the potential of emerging functional materials—including perovskites, MXenes, and quantum dots. In the domain of thermal expansion materials, which serve as the acoustic engine of the device, we analyzed the transition from rigid polymers to high-performance elastomers. PDMS has established itself as the benchmark material owing to its high coefficient of thermal expansion and mechanical flexibility. We also discussed cutting-edge innovations, such as self-healing PDMS based on metal–ligand coordination for enhanced durability, and ultra-soft elastomers like Ecoflex with low Young's modulus. In conclusion, the advancement of LGUS technology is intrinsically linked to the synergistic engineering of these composite materials. These material innovations are poised to drive LGUS devices from laboratory research into widespread diagnostic and therapeutic practice.

References

1 Li, J., Ma, Y., Zhang, T., Shung, K. K., and Zhu, B. (2022). Recent advancements in ultrasound transducer: from material strategies to biomedical applications. *BME Frontiers*, *2022*, 9764501.

2 Chen, S.-L. (2016). Review of laser-generated ultrasound transmitters and their applications to all-optical ultrasound transducers and imaging. *Applied Sciences*, *7*(1), 25.

3 Di, J., Kim, J., Hu, Q., Jiang, X., and Gu, Z. (2015). Spatiotemporal drug delivery using laser-generated-focused ultrasound system. *Journal of Controlled Release*, *220*, 592–599.

4 Lan, L., Xia, Y., Li, R., Liu, K., Mai, J., Medley, J. A., and Cheng, J.-X. (2018). A fiber optoacoustic guide with augmented reality for precision breast-conserving surgery. *Light: Science & Applications*, *7*(1), 2. doi:10.1038/s41377-018-0006-0

5 Liu, S., Kim, H., Huang, W., Chang, W.-Y., Jiang, X., and Ryu, J. E. (2022). Multiscale and multiphysics FEA simulation and materials optimization for laser ultrasound transducers. *Materials Today Communications, 31*, 103599.

6 Hsieh, B.-Y., Kim, J., Zhu, J., Li, S., Zhang, X., and Jiang, X. (2015). A laser ultrasound transducer using carbon nanofibers–polydimethylsiloxane composite thin film. *Applied Physics Letters, 106*(2), 021902.

7 Zhang, T., Li, C.-H., Li, W., Wang, Z., Gu, Z., Li, J., and Zhu, B. (2024). A self-healing optoacoustic patch with high damage threshold and conversion efficiency for biomedical applications. *Nano-Micro Letters, 16*(1), 122.

8 Yang, D. J., Wang, S. G., Zhang, Q., Sellin, P. J., and Chen, G. (2004). Thermal and electrical transport in multi-walled carbon nanotubes. *Physics Letters A, 329*(3), 207–213.

9 Das, H. (2017). Thermal transport in multi-walled carbon nanotubes. *Research & Reviews: Journal of Material Sciences, 5*(3), 423.

10 Masarapu, C., Henry, L. L., and Wei, B. (2005). Specific heat of aligned multiwalled carbon nanotubes. *Nanotechnology, 16*(9), 1490–1494.

11 Du, X., Li, J., Niu, G., Yuan, J.-H., Xue, K.-H., Xia, M., and Tang, J. (2021). Lead halide perovskite for efficient optoacoustic conversion and application toward high-resolution ultrasound imaging. *Nature Communications, 12*(1), 3348.

12 Colchester, R. J., Alles, E. J., and Desjardins, A. E. (2019). A directional fibre optic ultrasound transmitter based on a reduced graphene oxide and polydimethylsiloxane composite. *Applied Physics Letters, 114*(11), 113505.

13 Buma, T., Spisar, M., and O'Donnell, M. (1999). Thermoelastic generation of ultrasound using an erbium doped fiber amplifier. *IEEE Ultrasonics Symposium, 2*, 1253–1256.

14 Yang, H., Kim, J. S., Ashkenazi, S., O'Donnell, M., and Guo, L. J. (2006). Optical generation of high frequency ultrasound using two-dimensional gold nanostructure. *Applied Physics Letters, 89*(9), 093901.

15 Hou, Y., Kim, J. S., Ashkenazi, S., Huang, S. W., Guo, L. J., and O'Donnell, M. (2007). Broadband all-optical ultrasound transducers. *Applied Physics Letters, 91*(7), 073507.

16 Lee, T., and Guo, L. J. (2017). Highly Efficient Photoacoustic Conversion by Facilitated Heat Transfer in Ultrathin Metal Film Sandwiched by Polymer Layers. *Advanced Optical Materials, 5*(2), 1600421.

17 Kang, S., Yoon, Y., Kim, J., and Kim, W. (2013). Thermoelastic response of thin metal films and their adjacent materials. *Applied Physics Letters, 102*(2), 021908.

18 Chang, W. Y., Huang, W. B., Kim, J., Li, S. B., and Jiang, X. N. (2015). Candle soot nanoparticles-polydimethylsiloxane composites for laser ultrasound transducers. *Applied Physics Letters*, *107*(16), 161903.

19 Chen, Z., Wu, Y., Yang, Y., Li, J., Xie, B., Li, X., and Zhu, B. (2018). Multilayered carbon nanotube yarn based optoacoustic transducer with high energy conversion efficiency for ultrasound application. *Nano Energy*, *46*, 314–321.

20 Silva, A. D., Henriques, C. A., Malva, D. V., Calvete, M. J. F., Pereira, M. M., Serpa, C., and Arnaut, L. G. (2020). Photoacoustic generation of intense and broadband ultrasound pulses with functionalized carbon nanotubes. *Nanoscale*, *12*(40), 20831–20839.

21 Jiang, Y., Lee, H. J., Lan, L., Tseng, H.-A., Yang, C., Man, H.-Y., and Cheng, J.-X. (2020). Optoacoustic brain stimulation at submillimeter spatial precision. *Nature Communications*, *11*(1), 881.

22 Vella, D., Pereira, D. A., Mrzel, A., Vengust, D., Drnovšek, A., Arnaut, L. G., and Jezeršek, M. (2024). Picosecond photoacoustic generation of ultrasounds with composites of graphene-decorated gold nanoparticles. *Nano Energy*, *131*, 110236.

23 Yadav, D., Amini, F., and Ehrmann, A. (2020). Recent advances in carbon nanofibers and their applications—A review. *European Polymer Journal*, *138*, 109963.

24 Huang, W., Chang, W. Y., Kim, J., Li, S., Huang, S., and Jiang, X. (2016). A novel laser ultrasound transducer using candle soot carbon nanoparticles. *IEEE Transactions on Nanotechnology*, *15*(3), 395–401.

25 Chang, W. Y., Zhang, X. A., Kim, J., Huang, W., Bagal, A., Chang, C. H., and Jiang, X. (2018). Evaluation of photoacoustic transduction efficiency of candle soot nanocomposite transmitters. *IEEE Transactions on Nanotechnology*, *17*(5), 985–993.

26 Baac, H. W., Ok, J. G., Lee, T., and Jay Guo, L. (2015). Nano-structural characteristics of carbon nanotube–polymer composite films for high-amplitude optoacoustic generation. *Nanoscale*, *7*(34), 14460–14468.

27 Li, J., Xu, J., Liu, X., Zhang, T., Lei, S., Jiang, L., and Zhu, B. (2020). A novel CNTs array-PDMS composite with anisotropic thermal conductivity for optoacoustic transducer applications. *Composites Part B: Engineering*, *196*, 108073.

28 Kwon, S. W., Choi, W. Y., Jo, H. G., and Park, K. K. (2020). Frequency modulation of laser ultrasound transducer using carbon nanotube-coated polyethylene microsphere. *Journal of the Acoustical Society of America*, *147*(4), EL351–EL356.

29 Noimark, S., Colchester, R. J., Blackburn, B. J., Zhang, E. Z., Alles, E. J., Ourselin, S., and Desjardins, A. E. (2016). Carbon-nanotube–PDMS composite coatings on optical fibers for all-optical ultrasound imaging. *Advanced Functional Materials, 26*(46), 8390–8396.

30 Colchester, R. J., Noimark, S., Zhang, E. Z., Beard, P. C., and Desjardins, A. E. (2018, 22–25 Oct. 2018). *Dual-Modality All-Optical Ultrasound and Photoacoustic Imaging Using Permanent Marker Ink*. Paper presented at the 2018 IEEE International Ultrasonics Symposium (IUS).

31 Bodian, S., Colchester, R. J., Macdonald, T. J., Ambroz, F., Briceno de Gutierrez, M., Mathews, S. J., and Noimark, S. (2021). CuInS2 quantum dot and polydimethylsiloxane nanocomposites for all-optical ultrasound and photoacoustic imaging. *Advanced Materials Interfaces, 8*(20), 2100518.

32 Colchester, R. J., Little, C. D., Alles, E. J., and Desjardins, A. E. (2021). Flexible and directional fibre optic ultrasound transmitters using photostable dyes. *OSA Continuum, 4*(9), 2488–2495.

33 Wu, H., Guan, Z., Ke, Y., Yu, X., Zhang, Z., Li, M., and Lu, H. (2023). MXene-based photoacoustic transducer with a high-energy conversion efficiency. *Optics Letters, 48*(21), 5563–5566.

34 Barbosa, R. C., and Mendes, P. M. (2022). A comprehensive review on photoacoustic-based devices for biomedical applications. *Sensors, 22*(23), 9541.

35 Ali, U., Abd, K. K. J. B., and Buang, N. A. (2015). A review of the properties and applications of poly (methyl methacrylate) (PMMA). *Polymer Reviews, 55*(4), 678–705.

36 Wang, L., Zhao, Y., Zheng, B., Huo, Y., Fan, Y., Ma, D., and Wang, P. (2023). Ultrawide-bandwidth high-resolution all-optical intravascular ultrasound using miniaturized photoacoustic transducer. *Science Advances, 9*(23), eadg8600.

37 Noimark, S., Colchester, R. J., Poduval, R. K., Maneas, E., Alles, E. J., Zhao, T., and Desjardins, A. E. (2018). Polydimethylsiloxane composites for optical ultrasound generation and multimodality imaging. *Advanced Functional Materials, 28*(9), 1704919.

38 Li, C.-H., Wang, C., Keplinger, C., Zuo, J.-L., Jin, L., Sun, Y., and Bao, Z. (2016). A highly stretchable autonomous self-healing elastomer. *Nature Chemistry, 8*(6), 618–624.

39 Na, H., Park, J., & Jeong, K. H. (2023). Plasmon-enhanced optoacoustic transducer with Ecoflex thin film for broadband ultrasound generation using overdriven pulsed laser diode. Journal of Biomedical Optics, *28*(12), 125005–125005.

4

All-optical Ultrasound Imaging

All-optical ultrasound (AOUS) refers to a technique in which the generation and reception of ultrasound are solely accomplished by optical means. Two separate optical processes underlie this paradigm's operation: (i) Ultrasound generation: pulsed laser irradiation of photoacoustic composite materials produces ultrasound by converting light energy into mechanical vibrations through the photoacoustic effect; (ii) Ultrasound reception: the echo reception is achieved through interferometric detection of strain-induced optical modulation. The fundamental element for generating ultrasound is the laser-generated ultrasound (LGUS) device, which is usually built on transparent substrate surfaces or optical fiber interfaces, as described in Chapter 3. Similarly, ultrasound detection methods employ optical resonant structures such as the microring resonator (MRR) [1, 2], the Fabry–Pérot (F–P) resonator [3–5], the fiber-optic Bragg grating [6–8], and the Mach–Zehnder interferometer [9, 10]

AOUS imaging technology relies on variations in acoustic impedance within biological tissues. High-frequency sound waves propagate through tissues, experiencing attenuation and generating reflections (echoes) at interfaces between media with differing acoustic impedances. These echoes reveal the acoustic properties of tissues, highlighting structural differences in fat, muscle, organs, blood vessels, and bone. Depending on the application scenario (e.g., intravascular imaging [11, 12] and minimally invasive surgical guidance [4, 13–15]), the region of interest can be visualized as 2D or 3D images using specialized schemes like synthetic aperture processing, which are then displayed in real time. Due to their inherent benefits—including wide bandwidth (>20 MHz), miniature form factor, high sensitivity (low noise-equivalent pressure), affordability, and immunity to electromagnetic interference (EMI), AOUS systems have attracted increasing scholarly attention [7, 16, 17], becoming a viable alternative to traditional piezoelectric ultrasound in specific biomedical applications.

4.1 Optical Ultrasound Generation

The LGUS device is the key element for producing ultrasound waves in AOUS technology. It utilizes the photoacoustic effect to transform pulsed laser light into thermal energy, where the subsequent thermoelastic effect induces stress transients in the composite material to emit ultrasound waves. LGUS devices can be categorized into two types based on the laser excitation method as shown in Figure 4.1: (i) free-space light excitation: a laser generates a collimated, millimeter-diameter free-space beam, which directly irradiates the photoacoustic composite material without fiber coupling to emit ultrasound. (ii) Optical fiber excitation: laser pulses are coupled into an optical fiber via a lens. A photoacoustic composite coating fabricated at the distal end of the fiber generates a narrower acoustic beam upon excitation.

Given that photoacoustic materials are susceptible to damage from high-energy laser radiation, the choice between free-space light and optical

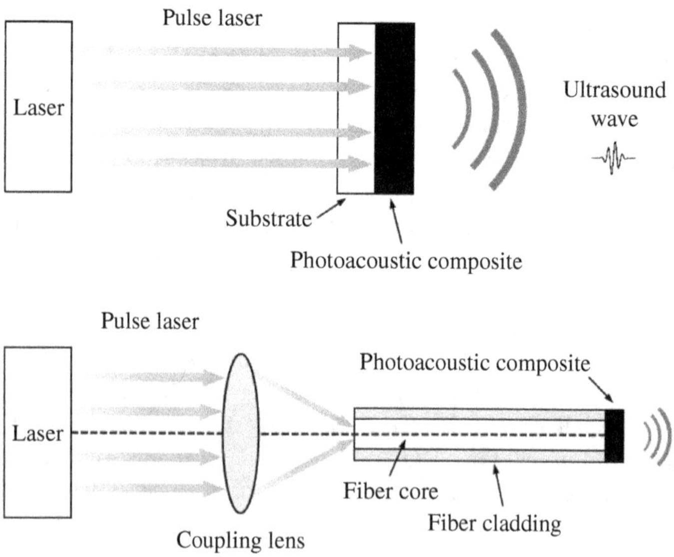

Figure 4.1 Schematic of ultrasound generation by free-space excitation and optical-fiber excitation.

fiber excitation is heavily influenced by the device's laser-induced damage threshold (LIDT). The LIDT, defined as the laser energy density (fluence) at which surface damage occurs, sets the critical upper limit for incident laser energy and is often correlated with the thickness of the photoacoustic composite. Compared to free-space configurations, fiber-optic LGUS devices typically possess a smaller active area and, consequently, a lower tolerance for high laser energy. In fiber-optic setups, the optical beam is highly focused on the coating, resulting in high local fluence. Moreover, heat accumulation from laser pulses can pose a risk of thermal damage; thus, pulse parameters must be carefully optimized to balance acoustic generation efficiency with the device's working life. Conversely, free-space light irradiation typically generates a millimeter-scale acoustic beam, providing a broader field of view. The larger optical spot creates a broad acoustic wavefront capable of capturing richer structural details. Additionally, fiber-generated ultrasound typically acts as a miniaturized point source (or focused beam), offering high spatial precision and flexibility due to the compact integration of optical fibers.

4.1.1 Free-space Excitation

Ultrasound generated by free-space light excitation typically forms millimeter-scale acoustic beams with a relatively wide field of view. Generally, two primary techniques are employed to fabricate photoacoustic composite coatings on optical ultrasound sensors or optical substrates: (i) transfer method: following the protocols outlined in Chapter 3, photoacoustic composite films are first fabricated. These films are detached from the glass substrate and immersed in ethanol. Once positioned over the interferometer head, the ethanol evaporates, promoting tight adhesion of the film to the sensor surface. (ii) Direct coating: the device surface is directly coated with a homogenous mixture of the thermal expansion material and the light-absorbing agent, followed by curing. To enhance mechanical robustness, the surface can be further encapsulated with pristine polydimethylsiloxane (PDMS) or other highly elastic polymers.

Ultrasound generated by free-space light typically forms millimeter-scale acoustic beams which are characterized by a wider field of view. Two main techniques are commonly used to deposit photoacoustic composite coatings onto transparent ultrasound sensors or optical structures:

i) Using the method outlined in Chapter 3, photoacoustic composite films with a thickness of roughly 60–100 μm were created [5]. The films were separated from the glass substrate and submerged in ethanol to swell them slightly. Once the films were positioned above the interferometer head, the ethanol volatilized, which caused them to contract tightly to the sensor's surface.

ii) The device's surface was evenly coated with a mixture of thermal expansion material and light-absorbing substance, which was subsequently cured. To improve the mechanical robustness of the LGUS device, the surface can also be encapsulated with PDMS or other highly elastic polymers [18].

4.1.2 Optical Fiber Excitation

The previous section discussed photoacoustic composites primarily developed for planar surfaces. However, fabricating these microscale coatings on optical fiber tips presents unique challenges; conventional thin-film deposition processes like spin-coating and chemical vapor deposition (CVD) are often unsuitable for such non-planar geometries.

Therefore, high-performance photoacoustic composites are typically deposited onto the fiber's distal end face using dip-coating techniques. The preparation strategies generally fall into two categories: (i) composite dip coating: the light-absorbing material (e.g., carbon nanomaterials) is thoroughly dispersed within the thermally expandable precursor (typically PDMS) to ensure homogeneity. The fiber's distal end is then dipped into this mixture, withdrawn, and cured to form an integrated photoacoustic coating. (ii) Double-layer coating: this scheme involves first depositing a concentrated absorber layer (e.g., a carbon-based organogel) onto the fiber tip. Subsequently, this active layer is encapsulated with an outer layer of pure PDMS to serve as the thermal expansion medium and protective barrier.

4.2 Optical Ultrasound Reception Technology

The operational bandwidth of conventional piezoelectric ultrasound transmitters and receivers is typically constrained (fractional bandwidth often <100%) in classic imaging modes, which hinders the advancement of ultrasound image quality. In contrast, the emission principle of LGUS differs fundamentally, allowing for the generation of high-frequency signals with ultrawide bandwidths (>100%). However, utilizing piezoelectric receivers to detect these broadband LGUS signals is often inefficient due to bandwidth mismatches, meaning the full potential of LGUS is not

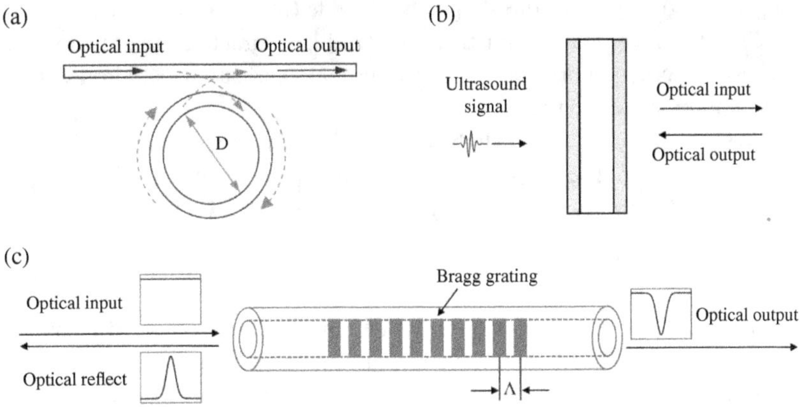

Figure 4.2 Optical detection technologies of ultrasound: (a) microring resonator (MRR); (b) Fabry–Pérot resonator; (c) fiber-optic Bragg grating (FBG).

completely realized. Consequently, optical ultrasound generation and detection are often employed in tandem to achieve bandwidth matching in all-optical ultrasound (AOUS) imaging.

Furthermore, the limited sensitivity of conventional piezoelectric sensors—particularly when miniaturized—remains a primary disadvantage. In contrast to piezoelectric sensors, optical sensors are virtually immune to EMI and offer significant advantages, including greater compactness, simpler beam focusing, and superior mechanical flexibility.

Optical ultrasound detection technology predates LGUS and has been utilized for decades in fields such as photoacoustic imaging and tomography. Polymer resonators are ideal optical ultrasound sensors due to their high sensitivity and acoustic impedance which is closely matched to that of biological tissues and water. This impedance matching minimizes acoustic reverberation and the parasitic effects of surface acoustic waves (SAWs), thereby preserving signal fidelity. As schematically shown in Figure 4.2, the primary polymer-based resonant sensors include Bragg gratings, MRRs, and Fabry–Pérot (F–P) sensors. Due to advances in miniaturization, these technologies hold the potential to replace piezoelectric detection arrays in high-resolution ultrasound imaging applications [1, 19].

4.2.1 Microring Resonator (MRR)

The MRR is a widely used ultrasound detection device in all-optical applications that leverages resonance effects at the micrometer scale. An MRR typically consists of a ring-shaped waveguide (resonant cavity) coupled to a linear bus waveguide, constructed from high-refractive-index materials. These devices are fabricated using various nanofabrication techniques, such as nanoimprint lithography.

Resonance occurs when light coupled into the microring satisfies the phase condition, meaning the optical path length of the ring is an integer multiple of the wavelength. The resonant wavelength, λ_m, is expressed as:

$$\lambda_m = \frac{\pi n_e D}{m} \tag{4.1}$$

where n_e is the effective refractive index inside the ring-shaped resonant cavity, D is the diameter of the ring-shaped resonant cavity, and m is the resonance order.

At resonance, light circulating in the loop interferes constructively (positive feedback), causing optical power to couple into or out of the ring depending on the configuration. Light wavelengths that do not meet the resonance condition pass through the bus waveguide unaffected. Ultrasound pressure alters the cross section and induces strain in the ring-shaped waveguide via the elasto-optic effect. This causes a variation in the effective refractive index and a shift in the resonant wavelength, which can be interrogated to measure the acoustic signal.

4.2.2 Fabry–Pérot Resonator

The Fabry–Pérot (F–P) resonator offers a broad bandwidth, high sensitivity, deep detection depth, and miniature acoustic element size, making it highly suitable for high-resolution AOUS imaging of biological tissues [20].

The F–P resonant cavity typically consists of a polymer film spacer sandwiched between two dichroic mirrors. Under normal conditions, a continuous-wave (CW) interrogation laser beam undergoes multiple reflections within the cavity. When the round-trip optical path length equals an integer multiple of the light's wavelength, optical resonance occurs. This results in a sharp drop in optical reflection (or a peak in transmission) near the resonance wavelength.

When ultrasound waves impinge upon the F–P cavity, they modulate the thickness of the polymer spacer and, consequently, the optical path length. By tuning the interrogation laser to the slope of the resonance curve, these path length variations are converted into intensity modulations of the reflected beam. The intensity of the ultrasound signal is then determined by demodulating this optical signal.

Polymers such as SU-8 or Parylene are frequently used for the spacer due to their high elasticity, which enhances sensitivity. Additionally, polymer materials minimize acoustic reflections at the sensor interface because they are acoustically impedance-matched to water or biological tissue. The mirrors are typically created by vacuum-depositing thin metallic or dielectric coatings on both sides of the spacer.

Despite its promise, the F–P resonator faces technical challenges in integrated LGUS array systems. First, fabricating an F–P cavity with uniform thickness across a large array aperture is critical to minimizing sensitivity variations between elements. Second, integrating high-density F–P-sensing elements differs significantly between systems designed for free-space light excitation versus those for optical fiber excitation.

4.2.3 Fiber-optic Bragg Grating (FBG)

FBG is another prominent optical resonant technology for ultrasound reception, known for its self-calibration and outstanding multiplexing capabilities. The fabrication of an FBG relies on the photosensitivity of the optical fiber core. A permanent periodic grating structure is inscribed by exposing the core transversely to an intense UV laser interference pattern. This exposure induces a periodic modulation of the core's refractive index. At these periodic index changes, a small fraction of light is reflected. These reflections interfere constructively only for a specific wavelength, known as the Bragg wavelength, λ_B, which is reflected back, while other wavelengths pass through with negligible attenuation. The Bragg wavelength is given by:

$$\lambda_B = 2n_{\text{eff}}\Lambda \tag{4.2}$$

where n_{eff} is the average effective refractive index of the grating region, and Λ is the grating period [6].

External factors such as strain, pressure, or temperature variations alter the grating period (Λ) and the refractive index (n_{eff}), causing a shift in λ_B. Since ultrasound is a mechanical wave, it induces dynamic stress changes in the FBG region, which modulates the Bragg wavelength; this is the fundamental principle of FBG ultrasound detection.

However, standard FBGs have limited sensitivity to high-frequency ultrasound due to their relatively long gauge lengths. To overcome this, π-phase-shifted FBG (π-FBG) sensors have been developed [21]. These are created by introducing a phase shift in the center of the periodic index modulation, analogous to a defect in a one-dimensional (1D) photonic crystal.

This phase shift alters the reflection and transmission spectra, creating a very narrow transmission peak (or reflection notch) within the broader stopband of the traditional FBG. This structure creates a high-Q cavity that significantly enhances sensitivity. Furthermore, by optically interrogating the sensor around the phase-shift region, the effective sensing length is reduced to a dimension smaller than the physical grating length. This inverse relationship between sensitivity and effective sensor length makes π-FBG sensors ideal for medical imaging applications requiring broadband, sensitive, and compact acoustic sensors.

4.3 All-Optical Ultrasound Imaging Systems

Ultrasound imaging remains the preeminent modality in medical diagnostics, celebrated for its noninvasive nature, real-time capability, and cost-effectiveness. The fidelity of ultrasound images is fundamentally governed by the system's vertical resolution, which theoretically correlates with the bandwidth of the ultrasound source $\frac{c}{2f\text{BW}}$. While high-frequency ultrasound offers superior resolution for visualizing fine lesion features, it traditionally suffers from rapid acoustic attenuation, severely limiting penetration depth and the ability to visualize deep biological structures. Consequently, traditional ultrasound technology faces an inherent trade-off: balancing high resolution with adequate imaging depth. For decades, piezoelectric transducers have been the standard hardware for generating and detecting ultrasound. However, the advancement of piezoelectric technology is increasingly constrained by two critical bottlenecks. First, the aforementioned trade-off between resolution and depth is difficult to circumvent with narrowband piezoelectric materials. Second, the requirement for complex electrical interconnects imposes physical limits on device miniaturization, hindering the development of ultra-compact probes for interventional applications.

AOUS technology has emerged as a transformative solution to these challenges. By leveraging the photoacoustic effect for generation and optical resonances for detection, AOUS devices can achieve ultrawide bandwidths exceeding 100%. This characteristic allows for the simultaneous achievement of high frequencies and broad bandwidths, effectively compensating for the loss of penetration depth typically associated with high-frequency imaging. Furthermore, AOUS systems offer immunity to EMI and facilitate seamless integration with optical fibers, enabling the creation of highly miniaturized probes.

These advantages have made AOUS highly attractive for clinical applications that require extreme miniaturization, such as intravascular imaging and minimally invasive surgical guidance. AOUS is also well suited for multimodal imaging environments, including MRI, because it is immune to EMI. The following sections provide a comprehensive review of the current state of AOUS imaging systems. We categorize these developments into two primary architectures based on their complexity and scanning mechanisms: single-element-based systems, which utilize mechanical scanning for high-resolution imaging, and array-based systems, which employ physical arrays to achieve visualization.

4.3.1 Single-Element-based Systems

Single-element-based AOUS systems represent the forefront of miniaturized imaging technology, garnering significant interest for their ability to deliver high-resolution diagnostics within an ultra-compact footprint. By eliminating the need for bulky piezoelectric arrays and complex electrical wiring, these systems integrate both LGUS emitters and optical fiber sensors into a singular, all-optical micro-probe. This architecture is particularly advantageous for interventional scenarios with stringent dimensional constraints, such as intravascular ultrasound (IVUS) and minimally invasive surgical guidance. To accommodate diverse anatomical targets and procedural requirements, the distal structures of these fiber-optic probes are engineered to support distinct propagation paths, primarily categorized into forward-viewing and side-viewing imaging configurations.

In forward-viewing imaging, the distal end of the ultrasound transmitting fiber is typically cleaved perpendicularly to its axis. The LGUS device is fabricated directly on this distal face, emitting ultrasonic waves along the probe's axial direction to visualize structures directly ahead. The research team led by Adrien E. Desjardins at the University College London (UCL) has been at the forefront of this field, demonstrating a progressive evolution of all-optical probes from benchtop prototypes to in vivo interventional tools.

Colchester et al. [22] first demonstrated an all-fiber ultrasound imaging probe utilizing a carbon nanotube (CNT)-PDMS composite for generation and a F–P resonator for detection. The CNT-PDMS composite was applied to the fiber tip via dip-coating and cured to form the ultrasound emission layer. This LGUS device achieved an acoustic pressure output greater than 4 MPa at the fiber face (200 μm core), with a center frequency of 18 MHz and a bandwidth of 15 MHz. For detection, the F–P optical resonator was constructed on a single-mode fiber, comprising a 20.1-μm epoxy resin spacer sandwiched between two high-reflectivity dichroic mirrors. Interrogated by

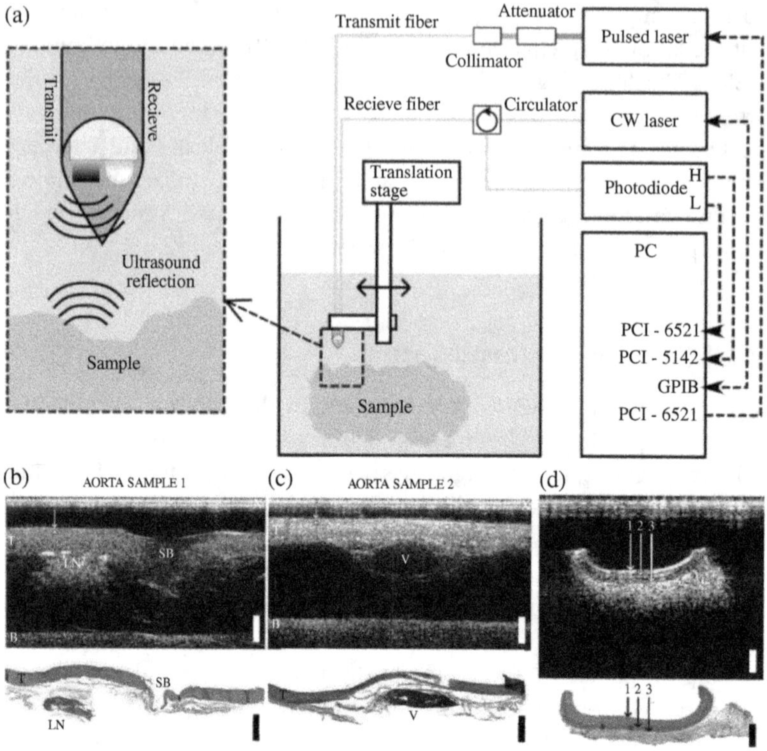

Figure 4.3 (a) Schematic of the imaging system and the probe structure. (b) Imaging of porcine aorta (Sample 1): Ultrasound image (top) and histology (bottom) showing tunica media (T), side branch (SB), lymph node (LN), tissue mount base (B), and crosstalk artifacts (X) [23] / Optica Publishing Group / CC BY 4.0. (c) Imaging of porcine aorta (Sample 2): Ultrasound (top) and histology (bottom) visualizing a vessel (V). Scale bars for (b) and (c) are 2 mm. (d) Imaging of porcine carotid artery: Ultrasound (top) and histology (bottom). Numbered arrows indicate layer boundaries: 1–2, tunica media; 2–3, external elastic lamina; 3, adventitia. Scale bar is 1 mm.

a tunable laser (1500–1600 nm), the F–P detector provided a receiving bandwidth of approximately 20 MHz with high sensitivity.

As shown in Figure 4.3(a), the emission and detection fibers were encapsulated within a stainless steel catheter (outer diameter <0.84 mm), creating a system suitable for minimally invasive interventions. The system demonstrated an optimal axial resolution of 58 μm (at 2.5 mm depth) and a lateral resolution of 88 μm (at 3.5 mm depth), with an effective imaging depth of 10 mm. By mechanically scanning the probe to synthesize a virtual array, the team successfully imaged porcine aorta and carotid arteries in vitro. As shown in Figure 4.3(b), (c), and (d), the aorta images clearly resolved features such as the tunica media, side branches, and lymph nodes. Notably, the carotid artery imaging clearly visualized its characteristic three-layer structure (tunica media, internal elastic lamina, and adventitia), showing high consistency with histological validation.

Subsequently, Alles et al. [23] proposed a highly directional all-optical "pencil beam" probe to address positioning inaccuracies and image artifacts common in synthetic aperture techniques. They compared a nondirectional probe (CNT-PDMS-coated fiber) with a focused pencil beam probe. As shown in Figure 4.4(a), the pencil beam probe utilized a concave lens (2 mm diameter, 3.4 mm radius of curvature) to collimate the acoustic beam. A large-core fiber (600 μm) guided light through an acrylic spacer to the lens surface, which was coated with carbon black as the photoacoustic absorber. The probe, encapsulated in heat-shrink tubing, generated a highly collimated beam with a width of less than 1 mm over a depth of 7 mm. The device exhibited a center frequency of ~20 MHz, a −6 dB bandwidth of 8–33 MHz, and an axial resolution of 75 μm. The lateral resolution was determined by the beam width (0.3 mm at 3 mm depth). Crucially, this design allowed single A-lines to be directly stitched into B-mode images without complex reconstruction. As shown in Figure 4.4(b), phantom imaging of a graphite target verified the system's robustness against positioning errors of up to ±300 μm. Ex vivo imaging of the porcine aorta (Figure 4.4(c, d)) clearly displayed the vascular wall and side branches in both 2D and 3D volumetric scans without reconstruction artifacts.

Furthermore, Finlay et al. [4] advanced this technology to in vivo applications by integrating the AOUS system into a transseptal puncture needle for real-time cardiac imaging. They utilized a multi-walled CNT (MWCNT)-PDMS composite on a 300 μm multimode fiber for emission and an F–P resonator on a single-mode fiber for detection. As shown in Figure 4.5(a), the fibers were housed within a custom needle (1.08 mm diameter), separated by a 0.1-mm metal septum to acoustically isolate the transmission and reception channels. The probe delivered a peak acoustic pressure of 8.8 MPa with a −6 dB bandwidth of 26.5 MHz. In preclinical swine studies, the system achieved an imaging depth greater than 2.5 cm with an axial resolution of 64 μm, enabling real-time visualization of cardiac structures to guide transseptal puncture. The system successfully identified key anatomical landmarks and visualized tissue displacement caused by heartbeats and respiration. As shown in Figure 4.5(b), M-mode imaging captured the rhythmic pulsation of the right atrial wall. Upon slight

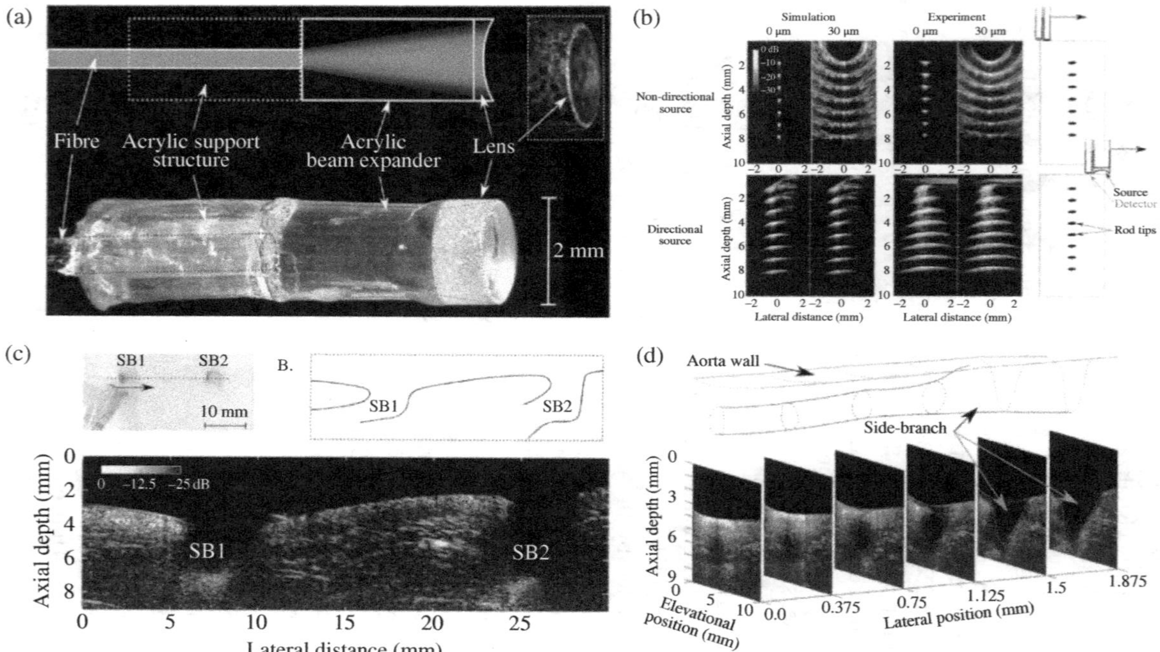

Figure 4.4 (a) Schematic and photograph of the focused optical ultrasound source. (b) (Top row) Simulated (first two columns) and measured (middle two columns) all-optical ultrasound images obtained from a phantom using an unfocused acoustic source. The phantom consisted of the tip of a single rod placed at different depths (right column); the resulting images are compounded into a single synthetic image. Simulations and experiments were performed both in the absence ("0 µm") and presence ("30 µm") of deliberate probe positioning errors. Positioning errors were applied to both the axial and the lateral axis and sampled from a uniform random distribution with a range of 30 µm. (Bottom row) The same panels are shown for the case in which all-optical pulse-echo ultrasound data were acquired using the focused acoustic source. (c) Photograph, schematic, and ultrasound image of the focused probe positioned above an ex vivo aortic section containing two side branches (SB1 and SB2). (d) Cross-sections through a volumetric all-optical ultrasound image of an aorta section containing a side branch [23] / Optica Publishing Group / CC BY 4.0.

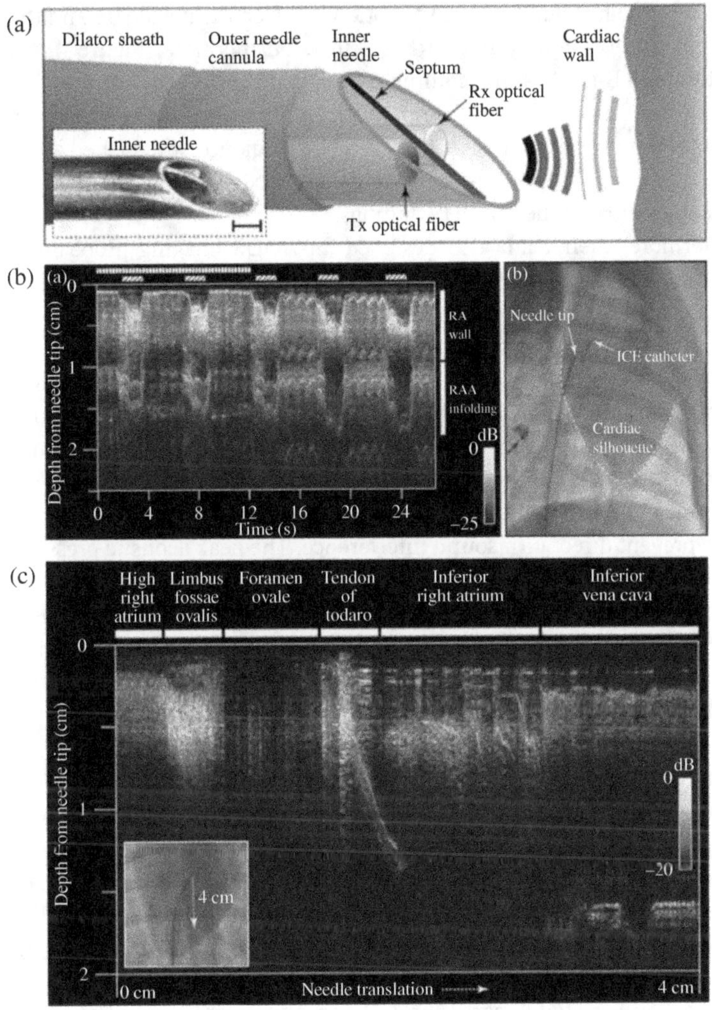

Figure 4.5 (a) Schematic and photo (inset) of the puncture needle probe. (b) All-optical ultrasound imaging from the right atrium (RA) with depth scans shown longitudinally in time for M-mode imaging (left). Conventional X-ray fluoroscopy imaging was performed concurrently (right). (c) Two-dimensional all-optical ultrasound imaging (B-mode) acquired during the manual translation of the needle tip across a distance of 4 cm. X-ray fluoroscopic imaging was acquired concurrently (inset) [4] / Springer Nature / CC BY 4.0.

retraction of the needle, the relative motion between the needle tip and the atrial wall (0.1–0.3 mm) was clearly distinct. Furthermore, Figure 4.5(c) demonstrates a B-mode scan acquired during manual pullback, visualizing the transition from the right atrium to the inferior vena cava, where the thin foramen ovale appeared as a hypoechoic region distinguished from the thickened atrial septum and the tendon of Todaro. Concurrent X-ray fluoroscopy validated these AOUS findings.

Furthermore, their team also designed an AOUS imaging probe integrated into a puncture needle and conducted in vivo experiments [4]. They fabricated a composite coating of MWCNT-PDMS at the distal end of 300 μm multimode fibers for ultrasound emission. The F–P resonator on a single-mode fiber is detected using a wavelength-tunable continuous-wave laser for ultrasound detection. As shown in Figure 4.5(a), two fibers are integrated into a custom puncture needle, which has a diameter of 1.08 mm, a length of 78 cm, and a 60° bevel at the distal end. The needle contains a 0.1 mm thick metal diaphragm, which is used to acoustically isolate the two fibers to prevent direct ultrasound interference. The peak acoustic pressure output of the probe can reach 8.8 MPa (1.5 mm away from the fiber), and the −6 dB bandwidth is 26.5 MHz. The imaging depth of the system is greater than 2.5 cm (in porcine heart tissue), and its axial resolution is 64 μm, which can be used for real-time visualization of heart wall movement and atrial structure. They used this system to guide transseptal puncture. In the live hearts, they observed the relative position of the needle tip and the heart tissue in real time through the imaging system, successfully identifying key anatomical structures. They could clearly observe the tissue displacement caused by the beating of the heart wall and breathing, avoiding accidental puncture of the distal wall or extracardiac structures. As shown in Figure 4.5(b), a deep scan was performed from the right atrium, and the M-mode imaging was displayed longitudinally over time. The needle tip points forward and is initially fixed on the wall of the right atrium (vertical line). There was a slight regression (>12 seconds), and the pulsation of the right atrial wall became obvious (the depth from the needle tip to the right atrial wall was 0.1–0.3 mm). When mechanical ventilation (oblique) is performed, the resulting cardiac displacement causes changes in the apparent thickness of the right atrial wall. Mural folding and movement of the right atrium can be seen outside the right atrial wall. Conventional X-ray fluoroscopy imaging was also carried out simultaneously. Further, as shown in Figure 4.5(c), two-dimensional full optical ultrasound imaging (B-mode) was obtained during the manual movement of the needle tip by 4 cm. As the needle tip moves from the lower part of the right atrium to the inferior vena cava, the thin foramen ovale appears as a hypoechoic area (with oblique artifacts of the intracardiac echocardiography catheter and sheath) between the thickened atrial septum and the Todaro tendon. X-ray fluoroscopy imaging was carried out simultaneously.

While the carbon-based composites represented a significant leap in miniaturized ultrasound generation, recent research has shifted toward novel functional materials to further break through the limits of energy conversion efficiency and acoustic output. Unlike the carbon-based composites that rely primarily on optimizing the concentration and dispersion of absorbers within a polymer matrix, new strategies focus on leveraging the intrinsic thermodynamic properties of semiconductor materials.

In this context, Du et al. [16] pioneered a forward-viewing AOUS system utilizing lead halide perovskite ($MAPbI_3$) as the photoacoustic generation medium. This material offers distinct advantages over traditional carbon nanomaterials, primarily due to its superior optical absorption and optimized thermal dynamics. Specifically, $MAPbI_3$ exhibits a high absorption coefficient of up to $6.7\,\mu m^{-1}$ (at 532 nm). Furthermore, the material possesses a remarkably low specific heat capacity (~308 $J\,kg^{-1}\,K^{-1}$—approximately 1/2 to 1/5 that of the carbon-based materials) and a small thermal diffusivity ($0.145\,mm^2\,s^{-1}$). Crucially, this low diffusivity is attributed to an energy overlap between optical and acoustic phonons, which induces the up-conversion of acoustic phonons and suppresses heat dissipation. This "phonon bottleneck" effect effectively localizes heat, dramatically enhancing the photoacoustic conversion efficiency.

Leveraging these intrinsic properties, the researchers fabricated a planar device with a PDMS/ $MAPbI_3$/PDMS sandwich structure (as shown in Figure 4.6(a)), where the perovskite layer (~323 nm) acts as the heater and the PDMS layers serve as thermal expansion and passivation barriers to prevent hydrolysis. Transitioning to a fiber-optic configuration, they dip coated the perovskite precursor (doped with 15 wt% PVP to suppress the coffee-ring effect) directly onto a 125-μm fiber tip, followed by a PDMS encapsulation layer. This perovskite-based LGUS device demonstrated a transformative improvement in acoustic performance compared to earlier CNT-based probes. Under nanosecond laser excitation (3 mJ), the device generated a peak acoustic pressure of 24.89 MPa, which is higher than the typical output reported for CNT-PDMS probes in similar configurations. The device also achieved a record photoacoustic conversion efficiency of 2.97×10^{-2} and a broad −6 dB bandwidth of 40.8 MHz. Additionally, by engineering the thickness of the PDMS expansion layer (5–30 μm), the

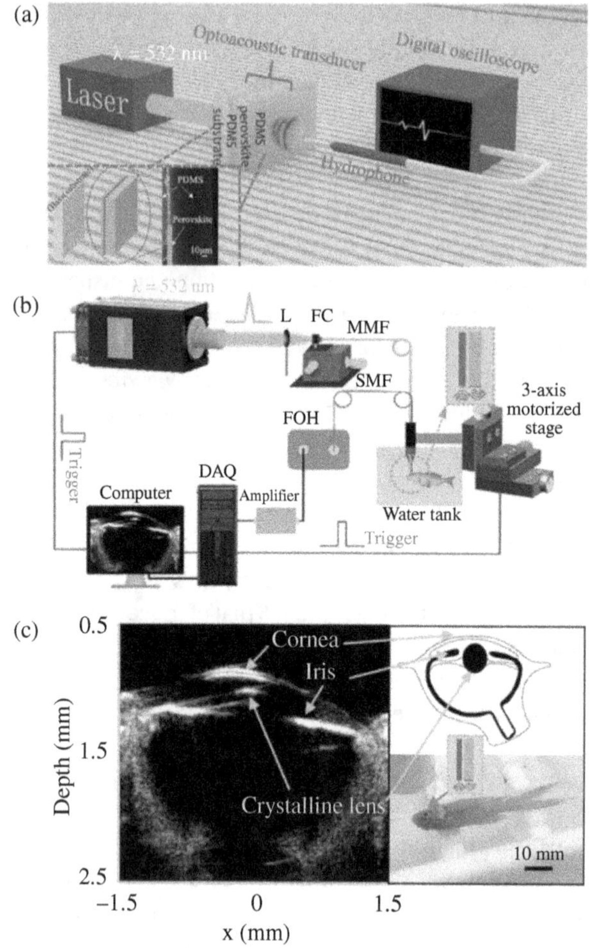

Figure 4.6 (a) Schematic of the LGUS device and the characterization system. The inset is the device structure and the cross-sectional scanning electron microscope (SEM) image of the device. (b) The all-optical ultrasound imaging system. (c) Ultrasound imaging of fisheye [16] / Springer Nature / CC BY 4.0.

ultrasonic waveform could be actively tuned between unipolar, bipolar, and multipolar modes. The resulting forward-viewing imaging system achieved an axial resolution of approximately 36 μm and demonstrated its capability through high-resolution ex vivo imaging of fish eyes, where it clearly resolved fine anatomical structures such as the cornea and lens surface as shown in Figure 4.6(b and c).

While the aforementioned forward-viewing systems demonstrated impressive material efficiency, they often relied on physically separated fibers for emission and detection. This discrete architecture introduces challenges such as alignment complexity, and increased probe diameter. To overcome these limitations, recent research has focused on monolithic integration, combining the ultrasound emitter and detector into a single fiber or chip to achieve coaxial alignment and extreme miniaturization.

Yang et al. [24] proposed a compact fiber-optic transducer that ingeniously utilizes the waveguide properties of a double-clad fiber (DCF) to isolate the optical paths of emission and detection within a single 125-μm strand. In this design, the pulsed excitation laser propagates through the large-diameter inner cladding, while the continuous-wave (CW) detection laser is confined to the single-mode core. The distal end of the fiber features a multilayered functional structure: a F–P cavity formed by two dichroic mirrors (highly reflective at 1550 nm, yet transparent to the excitation wavelength) sandwiches a polymer sensing film. This is followed by a PDMS isolation layer and, finally, the LGUS film prepared from a PDMS and carbon black mixture (mass ratio 5:1).

As shown in Figure 4.7(A), the excitation light passes through the transparent F–P sensors to strike the distal LGUS layer, generating ultrasound. Simultaneously, the detection light interrogates the F–P cavity, where returning ultrasound echoes modulate the polymer thickness and, consequently, the interference signal. The inclusion of the isolation layer is critical, as it shields the thermosensitive F–P cavity from the heat generated during optoacoustic conversion, ensuring signal stability. The resulting transducer demonstrated an output pressure of 0.45 MPa with a −6 dB bandwidth of 22 MHz. Crucially, it achieved a low noise equivalent pressure (NEP) of 72.8 Pa, indicating high sensitivity. In echo tests (Figure 4.7(B)) using an aluminum target, the device clearly resolved reflected signals. The authors also analyzed crosstalk artifacts (cross talks A and B), attributing them to the direct acoustic transmission from the generation layer to the sensor and reflections within the fiber coating, respectively. This highly integrated design offers a promising pathway for ultra-thin IVUS and endoscopic ultrasound (EUS) probes.

Taking integration a step further toward high-resolution-focused imaging, Nagli et al. [7] developed a transceiver utilizing silicon photonics technology. Unlike fiber-tip fabrication, this approach leverages a silicon-on-insulator (SOI) platform to create a robust, sub-millimeter probe with an aperture of less than 1 mm. The transducer comprises three key elements: a broadband silicon photonic detector based on a π-WBG, a micro-acoustic lens, and a

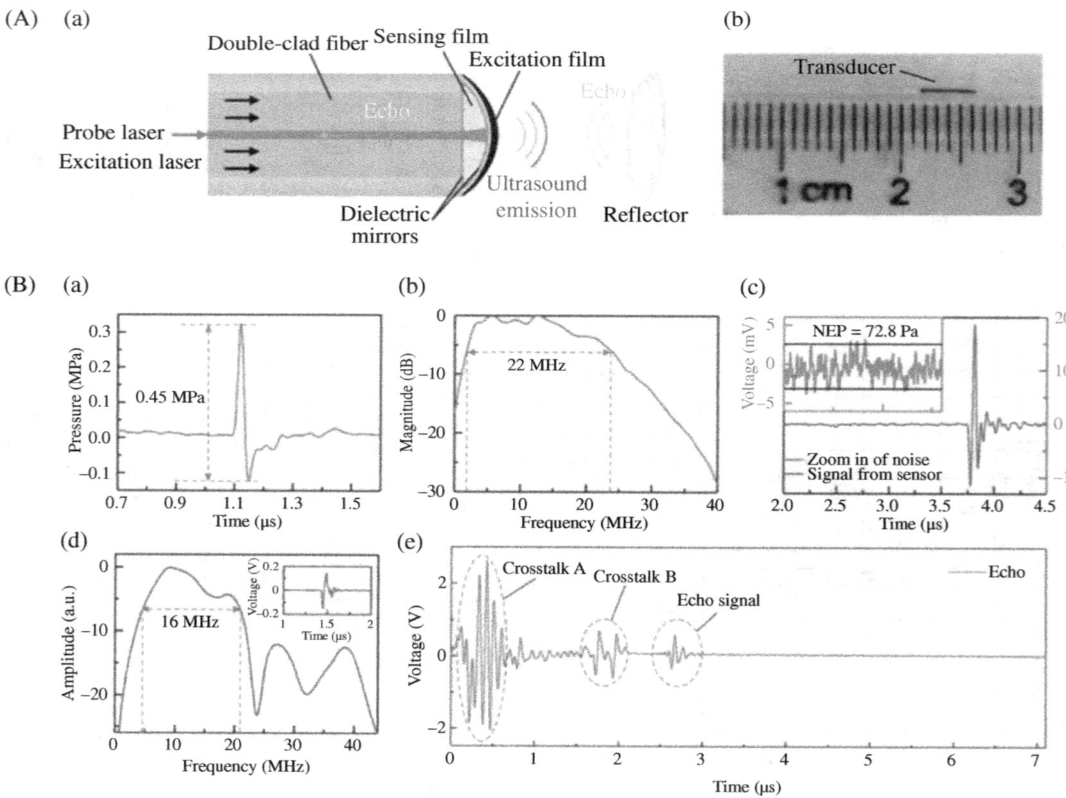

Figure 4.7 (A) The schematic (a) and photograph (b) of the fiber-optic ultrasound transducer. (B) The emission pressure (a) and bandwidth (b) of the transducer; the noise equivalent pressure [24] / with permission of Optica Publishing Group (c) and detection bandwidth (d) of the transducer; (e) the reflected signal from an aluminum plate.

metallic LGUS layer. As shown in Figure 4.8(a), a 40-nm chromium and 40-nm gold bilayer is coated onto the surface of an aspheric micro-lens.

Operationally, a 1-ns pulsed laser (1064 nm, ~10 μJ) transmitted via a multimode fiber irradiates the metallic layer on the lens. The generated ultrasound is collimated and focused by the lens into the tissue. Returning echoes are collected by the same lens and focused onto the highly sensitive π-WBG detector on the chip. This confocal design significantly enhances resolution.

Pulse-echo characterization using a steel plate revealed a center frequency of 33 MHz and a full width at half maximum (FWHM) bandwidth of 46 MHz. Most notably, the focusing capability yielded an axial resolution of 12 μm and a lateral resolution of 60 μm at the focal plane, significantly outperforming unfocused counterparts. Although the depth of field is relatively shallow (0.6 mm), this can be extended using synthetic aperture techniques. The system's capability was validated by imaging 50 μm tungsten wires (Figure 4.8(b)) and biological tissue (lamb meat) (Figure 4.8(c)), where coordinate transformation demonstrated its potential to generate IVUS-like images with exceptional structural detail.

While forward-viewing probes excel in guidance tasks (e.g., puncture), diagnostic imaging within luminal organs—such as blood vessels (IVUS) and the gastrointestinal tract (EUS)—necessitates a radial field of view to visualize cross-sectional tissue layers. Therefore, it is necessary to modify the structure at the distal end of the fiber, change the optical path to make it emit radially along the fiber, and fabricate an LGUS device in the direction of the laser emission to excite the device to emit ultrasonic waves radially, achieving side-viewing imaging.

Initial designs addressed the challenge of radial emission by mechanically altering the fiber tip. Colchester et al. developed an all-optical rotational B-mode imaging system utilizing a dual-fiber architecture [11]. As shown in Figure 4.9(A), the probe comprises a stationary omnidirectional detector (a concave Fabry–Pérot cavity with <100 Pa NEP) and a rotating emitter. The emitter fiber is polished to 45° and mirror coated to deflect light laterally onto a MWCNT-PDMS composite, creating a flat generation surface (~0.5 mm). This configuration produced a side-firing beam (1.87 MPa peak pressure, 31.3 MHz bandwidth) with a divergence of 13°. The axial resolution of the system is 50 μm. By mechanically rotating only the emission fiber, the system achieved imaging of porcine carotid arteries (ex vivo), achieving ultrasound imaging at a rate of 5 frames per second. As shown in Figure 4.9(B), the boundaries and wall structures of the vascular lumen can be clearly displayed in the ultrasound image. Meanwhile, by

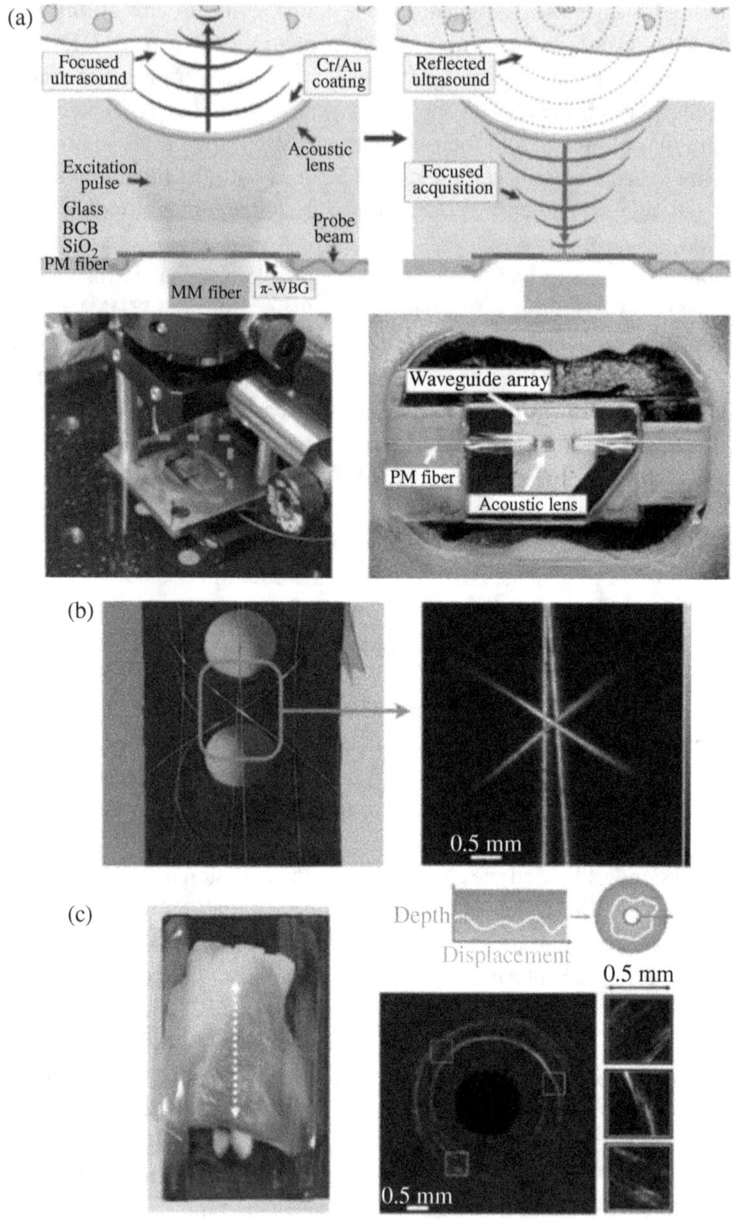

Figure 4.8 (a) Schematic (top) and photos (bottom) of the silicon-photonics-focused LGUS device. (b and c) Photos and ultrasound images of tungsten wires (b) and lamb meat tissue (c) [7] / with permission of Optica Publishing Group.

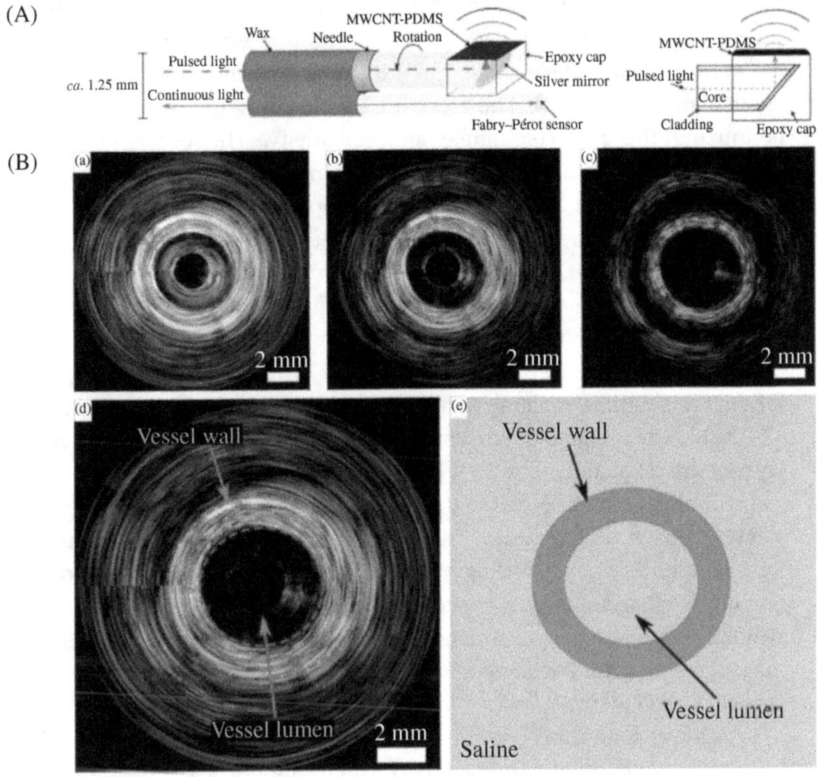

Figure 4.9 (A) Schematic of the side-viewing optical ultrasound probe. (B) Rotational optical ultrasound images of an ex vivo swine carotid artery with different frequency filters ((a) 0.5 MHz cut-off, (b) 10 MHz cut-off, (c) 20 MHz cut off, (d) depth-dependent filter (40 dB dynamic range)) and a schematic of the imaged vessel section (e) [11] / Springer Nature / CC BY 4.0.

using multifrequency filtering and depth-dependent frequency filtering, high frequencies can be used for high-resolution surface imaging and low frequencies for deep tissue penetration, which can effectively improve the image quality.

Translating this concept to clinical utility, Lewis-Thompson et al. advanced the design into a fully encapsulated endoscopic catheter suitable for the upper gastrointestinal tract [25]. Improvements included replacing MWCNT with reduced graphene oxide (rGO) for efficient generation and employed F–P fibers as ultrasound detectors. To achieve side-viewing, the distal end of a 200 μm core diameter fiber is inserted into a 0.5 mm × 0.5 mm square glass capillary tube. The end face is polished to 45° and coated with a silver mirror paint to form a lateral light reflection surface. The photoacoustic material is prepared on the side of the capillary tube as shown in Figure 4.10(A). The ultrasound emission fiber and the detection fiber are encapsulated in a torque coil (inner diameter 0.8 mm, outer diameter 1.2 mm). The receiving end extends approximately 1 mm, optimizing the receiving angle and sensitivity. The outer ultrasound transparent sheath tube (inner diameter 1.4 mm, outer diameter 1.6 mm) is compatible with the endoscopic channel. Ultrasound coupling is achieved by injecting physiological saline with a Y-type connector. The probe can be inserted through the standard endoscopic working channel (2 mm) and is compatible with clinical equipment (as shown in Figure 4.10(B)). Under the excitation of a 20 μJ laser, the probe can generate a peak acoustic pressure of 1.12 MPa, with a −6 dB bandwidth of up to 22 MHz, and has the potential for high-resolution imaging. The axial resolution of imaging system is 37–43 μm. The imaging depth can reach 20 mm, which can cover the entire layer of the esophagus. The SNR is up to 43 dB (in the shallow layer), and it can still remain above 30 dB in the deep layer. The imaging speed supports a fast pullback of 100 mm s^{-1}, and it only takes 0.5 seconds to image a 5 cm length. As shown in Figure 4.10(C), they conducted imaging of the porcine esophageal tissue and successfully distinguished the full-thickness structure of the esophagus, including the mucosa, submucosa, glands, and muscularis propria. The images were highly consistent with the histological structure.

To achieve long-time continuous rotational-scanning imaging in clinical applications, Zhang et al. developed an all-optical EUS system with strong system stability [26]. The imaging probe and system are shown in Figure 4.11(A). At the distal end of the ultrasound emission fiber in the imaging probe, a 532 nm nanosecond laser was used to irradiate the MWCNT-PDMS LGUS layer (located on the upper surface of the triangular prism), achieving lateral ultrasound emission. The peak acoustic pressure generated reached 1.5 MPa, and the bandwidth exceeded 30 MHz. At the ultrasound detector, the fiber laser sensor operates by utilizing the acoustic response of the fiber laser cavity (ultrasonic waves cause cavity shape changes, altering the laser frequency). The laser outputs two orthogonal polarization modes with a frequency difference of approximately 2 GHz. Ultrasonic waves cause a frequency shift of $\Delta\omega$, and the beat frequency signal changes to $2\Delta\omega$, enhancing the sensitivity. The ultrasound frequency response coefficient is 2.25 MHz kPa^{-1}, and the noise equivalent sound pressure (NEP) is as low as 37 Pa. It features high stability and strong resistance to thermal drift and mechanical interference. Two fibers are arranged in parallel and encapsulated in a steel catheter (with an outer

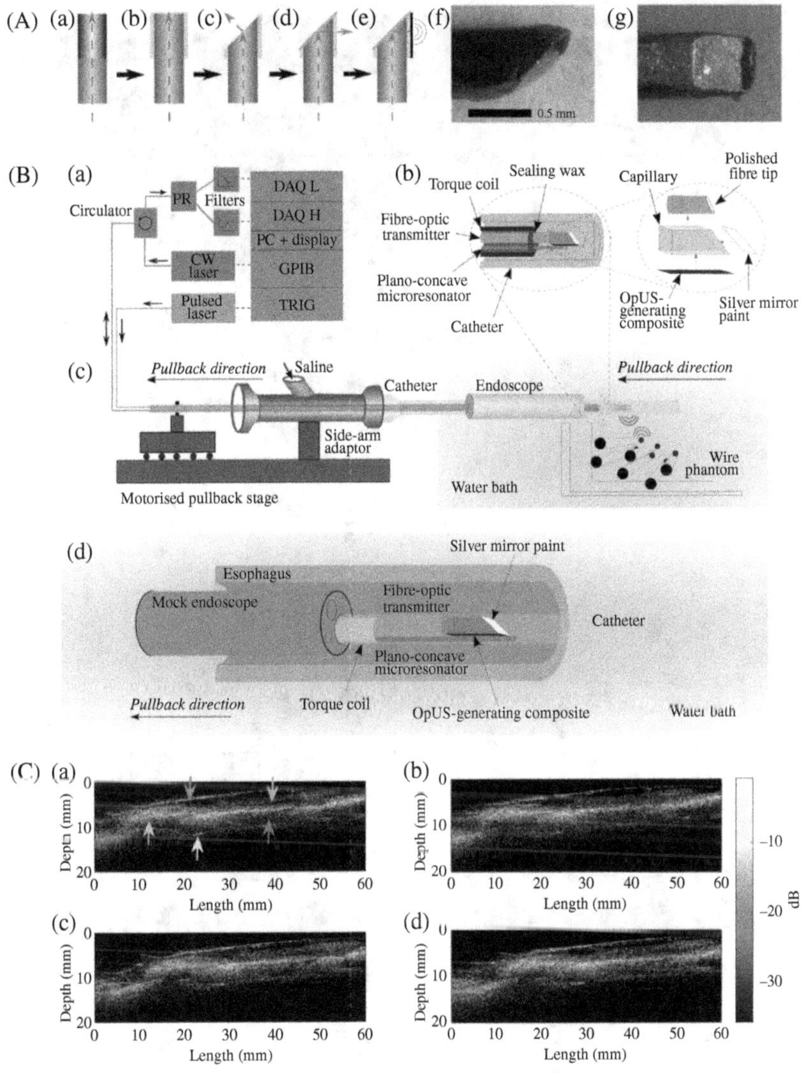

Figure 4.10 (A) Optical ultrasound transmitter fabrication process and stereo-microscope images of the lateral transmitter tip. (B) System for all-optical, side-viewing ultrasound imaging through fast pullback. (C) Reconstructed optical ultrasound image of esophageal tissue acquired using (top) 10 mm s^{-1} fast pullback reconstructed at an A-line spacing of (a) 25 µm and (b) 100 µm. (bottom) corresponding images taken using 100 mm s^{-1} fast-pullback at an A-line spacing of (c) 25 µm and (d) 100 µm. Inner edge of tissue (the downward arrow on the left), outer edge of tissue (the upward arrow in the middle), esophageal mucosa (the downward arrow on the right), submucosa with esophageal glands (the upward arrow on the left), and muscularis propria (the upward arrow on the right) [25] / Optica Publishing Group / CC BY 4.0.

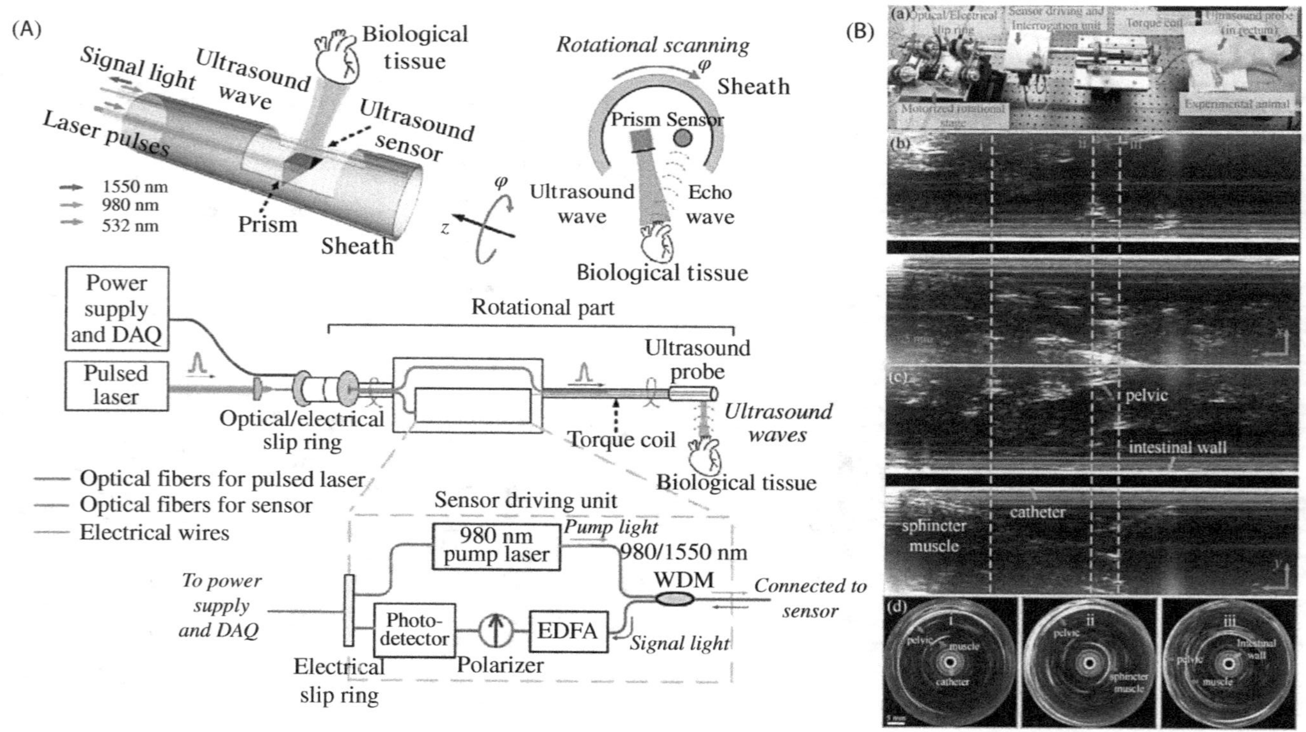

Figure 4.11 (A) Schematics of the ultrasound probe and the EUS system. (B) In vivo endoscopic imaging. (Top-bottom) Photograph of the EUS system during an in vivo animal experiment, imaging results over the $x-z$ and $y-z$ planes, three selected B-mode images at the position marked by the dashed lines [26] / with permission of Optica Publishing Group.

diameter of 2 mm and a length of 18 mm). In terms of rotation and signal transmission design, photoelectric/electrical signal slip rings are used to achieve single-channel photoelectric connection and support continuous rotation. The laser drive and signal acquisition unit are integrated at the rotating end, avoiding fiber entanglement. The electrical signal is transmitted to the control console through the slip ring, which enhances the system stability and imaging speed. The rotation is driven by the motor at a speed of 1 revolution per second, corresponding to a B-scan frame rate of 1 Hz. This system supports continuous pullback scanning at a speed of 50 $\mu m\ s^{-1}$ and a total travel of approximately 7 cm. The lateral resolution of the system is 0.97 mm (at a depth of 3 mm)—2.97 mm (at a depth of 9 mm), and the axial resolution is 192–240 μm (at a depth of 3 mm to 12 mm). In the in vivo imaging experiment of the rat rectum, they achieved imaging with a length of 7 cm and a volume of 4 cm × 4 cm × 6.8 cm, successfully identifying the inner wall of the rectal wall, the muscle layer, and the pelvic cavity (as shown in Figure 4.11(B)). The SNR is 27.8 dB at 5.97 mm and still remains 17.3 dB at 12.48 mm. The imaging depth and resolution are superior to those of traditional piezoelectric EUS probes.

To further improve resolution for intravascular applications, Wang et al. designed a high-resolution intravascular ultrasound probe using concave LGUS materials and π-FBG fiber sensors [12]. The LGUS material is MWCNT-PMMA (polymethylmethacrylate) composite film. After mixing MWCNT with PMMA, a concave film with a thickness of 5 μm and a curvature radius of 3 mm was formed through scraping, drying, and hot pressing, which was used to generate focused ultrasonic waves and thereby improve the lateral resolution. The ultrasound detector is a π-phase-shifted fiber Bragg grating (π-FBG), which is built on a single-mode fiber and features high sensitivity and broadband response capability. By adding a backing material (acoustic matching adhesive containing tungsten particles) to regulate its resonance mode, the frequency response range is significantly broadened, eliminating the response blind zone of traditional FBG in 50–60 MHz, and achieving a −6 dB bandwidth of 147%. The distal end of the ultrasound emission fiber is polished to 42°, guiding the sub-nanosecond laser (750 ps, 6 μJ) to the concave LGUS film to excite high-frequency broadband ultrasonic waves. An isolation plate is set between the excitation unit and the receiving unit to avoid direct acoustic interference. The catheter rotates and pulls back with the help of a motor, achieving 360° rotational imaging and automatic retraction. The center frequency of imaging system is 27.5 MHz, and the −6 dB bandwidth is 147% (7.3–47.7 MHz), which is much higher than that of traditional piezoelectric ultrasound transducers

(~40%). The axial resolution is 18.6 μm, the lateral resolution is 124 μm (at the focus), and the imaging depth can reach 7 mm.

Similar to the forward-viewing AOUS imaging, the side-viewing probe is also more inclined to be designed as an integrated transceiver. In particular, devices with separated emission and detection have a significant adverse impact on the movement of the probe in rotating side-viewing imaging. Therefore, the side-viewing probe with integrated emission and detection is more suitable for clinical applications. Richard J. Colchester et al. achieved a side-viewing AOUS transducer based on a single double-clad fiber [27]. The transverse dimension of the fiber ultrasound transducer is highly miniaturized (<0.8 mm). To simultaneously provide multimode light for ultrasound excitation and single-mode light for ultrasound detection, they connected a double-clad fiber with a small section of single-mode fiber. Multimode light couples out of the fiber at the connection point and propagates in the capillary, reflecting to the radial direction at the polished end face, which is used to excite the photoacoustic material coated on the side to generate ultrasonic waves. Single-mode light is transmitted to the planar concave microcavity resonator at the distal end of the single-mode fiber for detecting ultrasound signals. The device principle and preparation process are shown in Figure 4.12 (a and b). The LGUS film is an rGO/PDMS

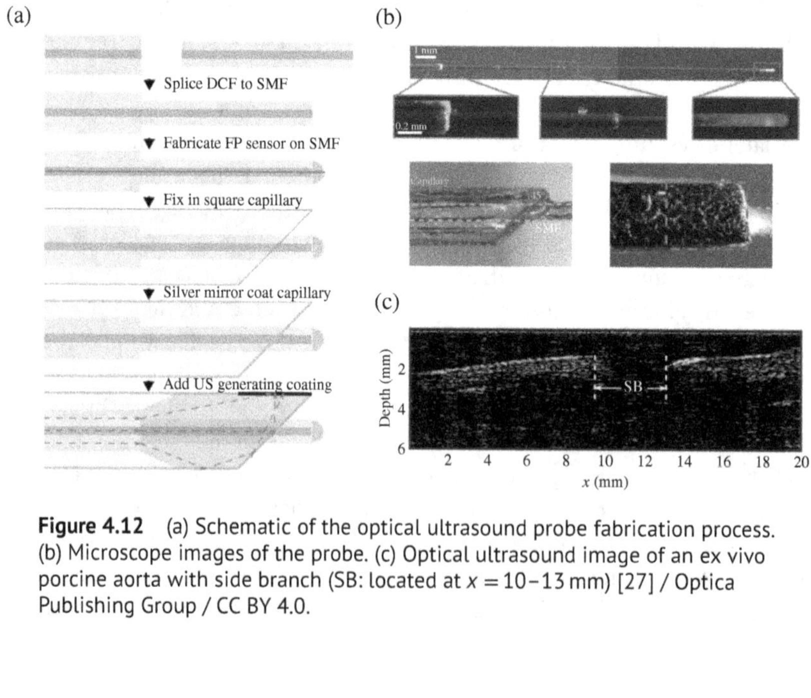

Figure 4.12 (a) Schematic of the optical ultrasound probe fabrication process. (b) Microscope images of the probe. (c) Optical ultrasound image of an ex vivo porcine aorta with side branch (SB: located at $x = 10$–13 mm) [27] / Optica Publishing Group / CC BY 4.0.

composite material. During preparation, rGO is dispersed in xylene, coated, and dried, and then coated with diluted PDMS and cured. The device can generate acoustic pressure exceeding 0.4 MPa, with a center frequency of approximately 15 MHz and a bandwidth greater than 27 MHz. The image resolution is approximately 50 μm in the axial direction and 190 μm in the lateral direction, with an imaging depth greater than 3 mm and an SNR of about 26 dB. They used the pullback method for imaging. The probe was shifted at a speed of 5 mm s^{-1}, and 800 A-lines (with an acquisition rate of 200 Hz) were collected within a scanning range of 20 mm. The total imaging time was approximately four seconds. Here, no image reconstruction is required. High-resolution images are directly obtained through processing such as filtering and Hilbert transform. Using this device for in vitro imaging of the porcine aorta, the vascular wall structure and collateral openings can be clearly distinguished. In the images, the vascular wall shows a bright boundary, with speckled echoes inside, and the collateral openings appear as signal-free areas, as shown in Figure 4.12(c). This study significantly enhances the integration and imaging efficiency of AOUS devices. Meanwhile, due to its side-viewing capability, they also provide a new technical reference for real-time high-resolution imaging in vascular and other lumen environments in the future.

4.3.2 Array-based Systems

As previously introduced, single-element AOUS systems can acquire high-quality 2D images through mechanical scanning. However, this approach relies on physically moving the transducer to "synthesize" an array, which inherently limits acquisition speed and fails to meet the core requirement of "real-time" performance for clinical diagnostics. Consequently, the development of array-based AOUS systems has become a pivotal research direction to overcome this bottleneck.

Erwin J. Alles et al. presented the first AOUS imaging system capable of video-rate real-time 2D imaging of biological tissues, addressing the longstanding limitation of prolonged acquisition times that had hindered the clinical translation of previous AOUS technologies [28]. As shown in Figure 4.13, the experimental setup utilized a spatially extended nanocomposite generator that produced uniform pressures of 0.977 ± 0.061 MPa and wide bandwidths of 27.1 ± 2.17 MHz, coupled with a highly sensitive fiber-optic receiver based on a F-P cavity (125 μm diameter, 80 MHz bandwidth). To optimize acoustic directivity and minimize signal decay, the system employed eccentric illumination via cylindrical optics, while a

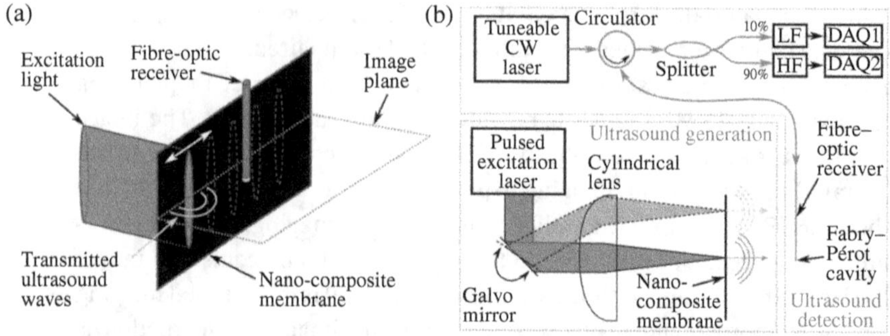

Figure 4.13 Schematic of the optical ultrasound generation and detection. (a) Using a cylindrical lens, excitation light was delivered to a small, eccentric area of an optically absorbing membrane comprising a nanocomposite. Ultrasound was generated photoacoustically in this area, and a linear acoustic aperture was scanned sequentially by translating the focal spot across the membrane using a galvo mirror. (b) Schematic of the setup used to optically generate and detect ultrasound [28] / Optica Publishing Group / CC BY 4.0.

fast galvanometer mirror scanning a 15.5-mm linear aperture enabled the dynamic synthesis of non-uniform, partially overlapping source arrays. By implementing both amplitude and source density apodization to reduce image artifacts, the system achieved a sustained frame rate of 15 Hz, a dynamic range of 30 dB, a penetration depth exceeding 6 mm, and a spatial resolution of 75 μm (axial) by 100 μm (lateral). The clinical relevance of the platform was validated through static ex vivo imaging of zebrafish anatomy and dynamic imaging of pulsatile ex vivo swine carotid arteries, the latter successfully capturing vessel wall deformation and the flow of glass microbubble contrast agents.

To address the clinical need for manually operable probes in ultrasound examinations, Erwin J. Alles et al. built on their prior research to create the first flexible, handheld AOUS imaging probe [29]. By integrating the versatility of fiber-optic single-element probes with the rapid video-rate imaging capability of benchtop systems that rely on scanning optics, this work represented a paradigm shift toward clinical translation.

As shown in Figure 4.14, the core design of this system involves using mirror galvanometers and a lens to couple pulsed excitation light into 64 discrete fibers arranged in a non-periodic pattern across a 25 mm linear aperture on an acrylic substrate. The distal ends of these fibers are coated with a MWCNT-PDMS composite film (49 ± 6 μm thick) that converts light to ultrasound via the photoacoustic effect, while a single centrally positioned fiber-optic detector with a plano-concave F–P cavity (125 μm

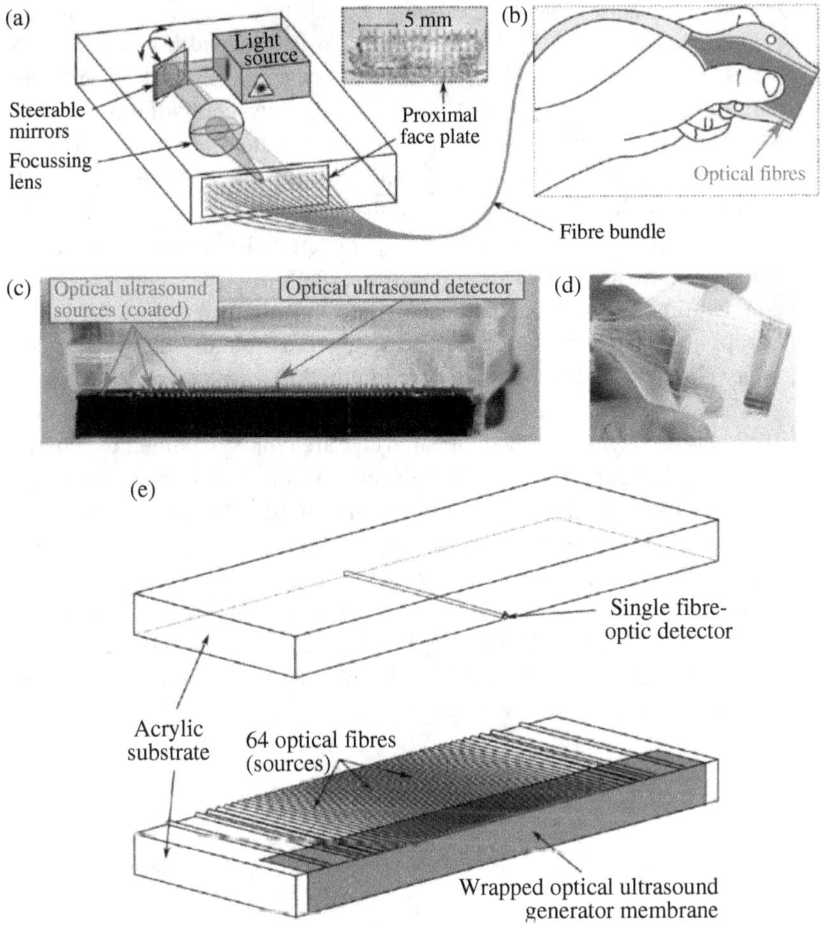

Figure 4.14 Schematic of the all-optical freehand and video-rate ultrasound imaging system. (a) Schematic of the all-optical ultrasound imaging setup. (b) Schematic of the distal end of the imaging probe, showing the tapered and aperiodic distribution of 64 grooves used to position and align the fiber-optic ultrasound sources. (c) Macro photograph of the all-optical ultrasound imaging aperture. (d) Photograph of the assembled distal end of the imaging probe. (e) Schematic (not to scale) of the distal end of the imaging probe [29] / with permission of Elsevier.

diameter) captures backscattered ultrasound signals. Signal processing employs a GPU-parallelized Delay & Sum algorithm for image reconstruction, achieving a sustained 11 Hz frame rate without signal averaging. Acoustical characterization at 1.6 mm from the probe showed uniform performance across the sources, with a peak pressure 0.24 ± 0.07 MPa, a center frequency of 11.2 ± 0.9 MHz, and a −6 dB bandwidth 17.4 ± 1.7 MHz (155% fractional bandwidth). To reduce image reconstruction artifacts, band-pass filtering (2–7 MHz) was applied. This configuration yielded a near-isotropic spatial resolution of 169 μm (lateral) × 173 μm (axial), a dynamic range of 30 dB, a SNR of 19 dB, and a signal-to-clutter ratio (SCR) of 21 dB. Experimental validations confirmed the system's efficacy: imaging a 27 μm tungsten wire point target demonstrated consistent resolution; dynamic imaging of a tissue-mimicking phantom visualized vessel walls and real-time needle insertion, albeit with 13 dB contrast and artifacts from needle echogenicity; and in-vivo imaging of a human common carotid artery clearly visualized the artery, monitored pulsatile diameter changes, and heart rate variations, with 13 dB contrast similar to phantom imaging.

As is generally understood, linear arrays are typically limited to generating 2D ultrasound images, whereas the acquisition of 3D ultrasound images traditionally necessitates the construction of a 2D array. Khoa Pham et al. developed a broadband all-optical plane-wave ultrasound imaging system for high-resolution 3D biological tissue imaging [30], addressing the limitations of conventional piezoelectric scanners (narrow bandwidth, large detection elements, limited 2D imaging) by integrating a planar F–P scanner for ultrasound detection and a CNT-PDMS composite film for ultrasound generation, as show in Figure 4.15.

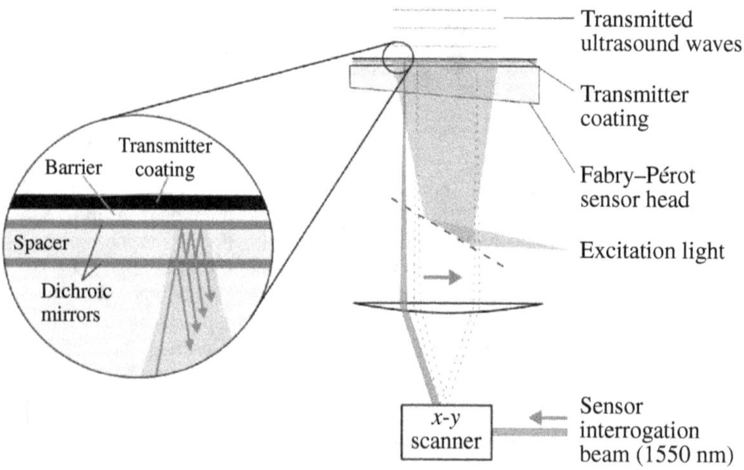

Figure 4.15 Schematic of the broadband all-optical plane-wave ultrasound imaging system based on a Fabry–Pérot scanner [30] / with permission of IEEE.

The core design centers on a planar F–P sensor head, comprising two dichroic mirrors separated by a 20-μm Parylene-C spacer and a 2.5-μm Parylene-C barrier layer. This sensor is coated with a 60–100 μm-thick CNT-PDMS composite film, which, upon excitation by nanosecond laser pulses, generates plane-wave ultrasound pulses. These pulses exhibit peak pressures in the MPa range (2.54 ± 0.24 MPa at 5 mm depth), a −6 dB bandwidth of 22 MHz, and wavefront planarity within λ/10 at 10 MHz. The scattered acoustic field is detected by raster-scanning a focused 1550 nm continuous-wave interrogation laser across the F–P sensor. The scan covers a 15 mm × 15 mm aperture with a 100 μm step size (22,500 spatial points). The F–P sensor demonstrates a −3 dB bandwidth of 30 MHz and an estimated noise-equivalent pressure (NEP) of ~0.3 kPa over 20 MHz.

Image reconstruction is performed using a k-space algorithm based on plane-wave decomposition and the inverse Fourier transform, following a 30 MHz low-pass filter. Postprocessing steps include Hilbert transform envelope detection, time-gain compensation, and Fourier interpolation to achieve a voxel size of 25 μm × 25 μm × 10 μm. System characterization revealed a depth-dependent lateral line-spread function (LSF) ranging from 46 to 65 μm (at depths of 1–12 mm) and a constant axial LSF of 44 μm across the field of view. Furthermore, thermal analysis confirmed a stable detection window of 50 μs (equivalent to an imaging depth of 37.5 mm) before thermally induced shifts in the interferometer transfer function (ITF) compromised sensitivity.

The system's high-fidelity 3D imaging performance was validated through a series of experiments. Complex geometries, including optical fiber knots and polymer leaf skeletons, were resolved with a resolution of approximately 50 μm. In tissue-mimicking phantoms designed with biologically relevant impedance contrasts, the system outperformed conventional 20 MHz and 14 MHz clinical probes, yielding higher axial resolution and finer speckle. Moreover, ex vivo imaging of porcine tissue and human lymph nodes revealed anatomical structures with exceptional clarity.

4.4 Summary

As a new type of ultrasound imaging technology, all-optical technology used for ultrasound excitation and detection enables the generation and reception of a wider bandwidth, which is expected to break through the bottleneck of traditional ultrasound imaging technology in terms of image quality, and has received widespread attention. However, since the acoustic performance and clinical performance of all-optical imaging technology are still far from that of commercial ultrasound imaging devices, the search for new materials with excellent performance and the optimization of the device structure in order to improve the imaging performance is the eternal theme of the development of all-optical imaging. At the same time, most of the imaging experimental objects in the field of scientific research are isolated animal tissues, and the feasibility and safety of the in vivo application of LGUS technology have not yet been systematically evaluated. AOUS imaging technology has a long way to go before it can be commercially applied and industrially mass-produced.

References

1. Huang, S. W., Chen, S. L., Ling, T., Maxwell, A., O'Donnell, M., Guo, L. J., and Ashkenazi, S. (2008). Low-noise wideband ultrasound detection using polymer microring resonators. *Applied Physics Letters, 92*(19), 193509.
2. Hsieh, B.-Y., Chen, S.-L., Ling, T., Guo, L. J., and Li, P.-C. (2014). All-optical scanhead for ultrasound and photoacoustic imaging—Imaging mode switching by dichroic filtering. *Photoacoustics, 2*(1), 39–46.
3. Ashkenazi, S., Hou, Y., Buma, T., and O'Donnell, M. (2005). Optoacoustic imaging using thin polymer étalon. *Applied Physics Letters, 86*(13), 134102.
4. Finlay, M. C., Mosse, C. A., Colchester, R. J., Noimark, S., Zhang, E. Z., Ourselin, S., and Desjardins, A. E. (2017). Through-needle all-optical ultrasound imaging in vivo: a preclinical swine study. *Light: Science & Applications, 6*(12), e17103–e17103.
5. Chen, Y., Wang, Y., Lv, T., Zhang, J., and Yu, H. (2023). Single optical fiber based forward-viewing all-optical ultrasound self-transceiving probe. *Optics Letters, 48*(4), 868–871.
6. Daniel, C. B., Graham, T., Brian, C., and Wieslaw, J. S. (2003). Acousto-ultrasonic sensing using fiber Bragg gratings. *Smart Materials and Structures, 12*(1), 122.
7. Nagli, M., Koch, J., Hazan, Y., Levi, A., Ternyak, O., Overmeyer, L., and Rosenthal, A. (2023). High-resolution silicon photonics focused ultrasound transducer with a sub-millimeter aperture. *Optics Letters, 48*(10), 2668–2671.

8 Li, L., Xue, X., Zhong, H., Wang, X., Hou, C., You, D., and Guo, T. (2024). Directional fiber-optic ultrasonic transceiver using tilted fiber Bragg gratings. *IEEE Transactions on Instrumentation and Measurement*, *73*, 1–8.

9 Ouyang, B., Li, Y., Kruidhof, M., Horsten, R., van Dongen, K. W. A., and Caro, J. (2019). On-chip silicon Mach–Zehnder interferometer sensor for ultrasound detection. *Optics Letters*, *44*(8), 1928–1931.

10 Bauer-Marschallinger, J., Felbermayer, K., and Berer, T. (2017). All-optical photoacoustic projection imaging. *Biomedical Optics Express*, *8*(9), 3938–3951. doi:10.1364/BOE.8.003938

11 Colchester, R. J., Little, C., Dwyer, G., Noimark, S., Alles, E. J., Zhang, E. Z., and Desjardins, A. E. (2019). All-optical rotational ultrasound imaging. *Scientific Reports*, *9*(1), 5576.

12 Wang, L., Zhao, Y., Zheng, B., Huo, Y., Fan, Y., Ma, D., and Wang, P. (2023). Ultrawide-bandwidth high-resolution all-optical intravascular ultrasound using miniaturized photoacoustic transducer. *Science Advances*, *9*(23), eadg8600.

13 Xia, W., Mosse, C. A., Colchester, R. J., Mari, J. M., Nikitichev, D. I., West, S. J., and Desjardins, A. E. (2015, 21 June 2015). *Fiber Optic Photoacoustic Probe with Ultrasonic Tracking for Guiding Minimally Invasive Procedures*. Paper presented at the Opto-Acoustic Methods and Applications in Biophotonics II, Munich.

14 Ansari, R., Zhang, E. Z., Desjardins, A. E., and Beard, P. C. (2020). Miniature all-optical flexible forward-viewing photoacoustic endoscopy probe for surgical guidance. *Optics Letters*, *45*(22), 6238–6241.

15 Mathews, S. J., Little, C., Zhang, E., Beard, P., Mastracci, T., Rakhit, R., and Desjardins, A. E. (2023). Bend-insensitive fiber optic ultrasonic tracking probe for cardiovascular interventions. *Medical Physics*, *50*(6), 3490–3497.

16 Du, X., Li, J., Niu, G., Yuan, J.-H., Xue, K.-H., Xia, M., and Tang, J. (2021). Lead halide perovskite for efficient optoacoustic conversion and application toward high-resolution ultrasound imaging. *Nature Communications*, *12*(1), 3348.

17 Noimark, S., Colchester, R. J., Blackburn, B. J., Zhang, E. Z., Alles, E. J., Ourselin, S., and Desjardins, A. E. (2016). Carbon-nanotube–PDMS composite coatings on optical fibers for all-optical ultrasound imaging. *Advanced Functional Materials*, *26*(46), 8390–8396.

18 Buma, T., Spisar, M., and Donnell, M. O. (2003). A high-frequency, 2-D array element using thermoelastic expansion in PDMS. *IEEE Transactions on Ultrasonics, Ferroelectrics, and Frequency Control*, *50*(9), 1161–1176.

19 Czuchnowski, J., and Prevedel, R. (2021). Adaptive optics enhanced sensitivity in Fabry-Pérot based photoacoustic tomography. *Photoacoustics*, *23*, 100276.

20 Zhang, E., Laufer, J., and Beard, P. (2008). Backward-mode multiwavelength photoacoustic scanner using a planar Fabry-Perot polymer film ultrasound sensor for high-resolution three-dimensional imaging of biological tissues. *Applied Optics*, *47*(4), 561–577.

21 Shen, H., Wang, Z., Wu, Y., and Yang, B. (2016). One-dimensional photonic crystals: fabrication, responsiveness and emerging applications in 3D construction. *RSC Advances*, *6*(6), 4505–4520.

22 Colchester, R. J., Zhang, E. Z., Mosse, C. A., Beard, P. C., Papakonstantinou, I., and Desjardins, A. E. (2015). Broadband miniature optical ultrasound probe for high resolution vascular tissue imaging. *Biomedical Optics Express*, *6*(4), 1502–1511.

23 Alles, E. J., Noimark, S., Zhang, E., Beard, P. C., and Desjardins, A. E. (2016). Pencil beam all-optical ultrasound imaging. *Biomedical Optics Express*, *7*(9), 3696–3704.

24 Yang, L., Chen, G., Xu, D., Wang, A., Dai, C., and Sun, Q. (2022, May). Compact fiber optic ultrasound transducer with integrated ultrasound emission and detection. In *2022 Conference on Lasers and Electro-Optics (CLEO)* (pp. 1–2). IEEE.

25 Lewis-Thompson, I., Zhang, E. Z., Beard, P. C., Desjardins, A. E., and Colchester, R. J. (2023). All-optical ultrasound catheter for rapid B-mode oesophageal imaging. *Biomedical Optics Express*, *14*(8), 4052–4064.

26 Zhang, Q., Wu, C., Long, K., Huang, C., Zhong, X., Bai, X., and Guan, B. O. (2023). in vivo endoscopic ultrasound imaging with a rotational-scanning, all-optical ultrasound probe. *Optics Letters*, *48*(7), 1926–1929.

27 Colchester, R. J., Zhang, E. Z., Beard, P. C., and Desjardins, A. E. (2022). High-resolution sub-millimetre diameter side-viewing all-optical ultrasound transducer based on a single dual-clad optical fibre. *Biomedical Optics Express*, *13*(7), 4047–4057.

28 Alles, E. J., Noimark, S., Maneas, E., Zhang, E. Z., Parkin, I. P., Beard, P. C., and Desjardins, A. E. (2018). Video-rate all-optical ultrasound imaging. *Biomedical Optics Express*, *9*(8), 3481–3494.

29 Alles, E. J., Mackle, E. C., Noimark, S., Zhang, E. Z., Beard, P. C., and Desjardins, A. E. (2021). Freehand and video-rate all-optical ultrasound imaging. *Ultrasonics*, *116*, 106514.

30 Pham, K., Noimark, S., Huynh, N., Zhang, E., Kuklis, F., Jaros, J., and Beard, P. (2020). Broadband all-optical plane-wave ultrasound imaging system based on a Fabry–Perot scanner. *IEEE Transactions on Ultrasonics, Ferroelectrics, and Frequency Control*, *68*(4), 1007–1016.

5

Laser-Generated Ultrasound Therapy

Laser-generated ultrasound (LGUS) technology has emerged as a cutting-edge research direction in biomedical studies, due to its advantages such as high spatiotemporal resolution, wide-frequency band response, and miniaturized structure, demonstrating significant value in fields like neuroscience exploration and targeted therapy mechanism research. This chapter focuses on the three core application directions of LGUS in biomedical research. Starting from the technical characteristics and research value of each direction, it systematically elaborates on the technical principles and application prospects of LGUS technology in neuromodulation, drug delivery, and acoustic cavitation. In the field of neuromodulation, LGUS can generate submillimeter-level focused sound fields, enabling precise regulation at the single-neuron level by activating mechanically sensitive ion channels. This technology, which requires no genetic modification and has an extremely low risk of thermal damage, provides an innovative technical solution for high spatiotemporal precision neural stimulation. In terms of drug delivery, LGUS acts on microparticles through acoustic radiation force, promoting their effective displacement, thereby achieving remote control of microfluidic particles or on-demand drug release from nano-drug delivery systems, offering a feasible path for the noninvasive and precise development of drug delivery. In tissue resection and ablation, as the sound pressure of LGUS rises to the cavitation threshold, the induced acoustic cavitation enables regionally controllable tissue cutting and ablation. This allows for high-precision-targeted interventions, while also offering an innovative research approach to optimize therapeutic strategies in areas such as tumor ablation and thrombolysis. Figure 5.1 presents a simplified schematic overview of the three research directions for application. These applications fully exemplify the multidimensional value of LGUS, thereby

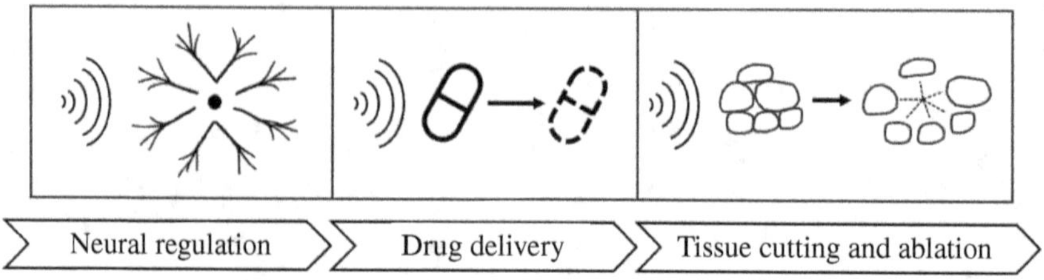

Figure 5.1 Schematic diagram of laser-generated ultrasound therapy.

furnishing robust support for investigations into precise neuromodulation and therapeutic interventions within the biomedical domain.

5.1 Neuromodulation

In neuroscience, neural stimulation has become a core method for understanding brain function and treating neurological disorders [1]. Traditional electrical stimulation typically requires implanted metal electrodes, which carry risks of surgical complications such as inflammation and bleeding. Noninvasive techniques, including transcranial magnetic stimulation (TMS) and transcranial electrical stimulation (tES), avoid invasive procedures but face limitations in spatial precision and penetration depth, making it difficult to accurately target specific brain regions. While optogenetics offers high spatial resolution and cell-type specificity, its reliance on viral transfection and genetic manipulation hinders clinical translation. Infrared neural stimulation, on the other hand, requires local temperature increases of several degrees Celsius to activate ion channels, raising significant safety concerns. In contrast, transcranial ultrasound stimulation emerges as a novel approach [2–5], which operates on the following principle: by focusing ultrasound waves onto specific brain regions, it modulates neuronal electrophysiological activity, neurotransmitter release, and neurovascular unit function via acoustic radiation force, ultimately achieving neural regulation and therapeutic intervention. This technology is characterized by its non-radiation, noninvasive, and real-time modulation capabilities, which help minimize the side effects and clinical risks associated with conventional treatments, demonstrating strong potential for clinical translation. As an emerging ultrasound modality, LGUS—compared to traditional piezoelectric ultrasound—features a simpler excitation mechanism and greater operational flexibility, offering distinct advantages in transcranial neuromodulation applications.

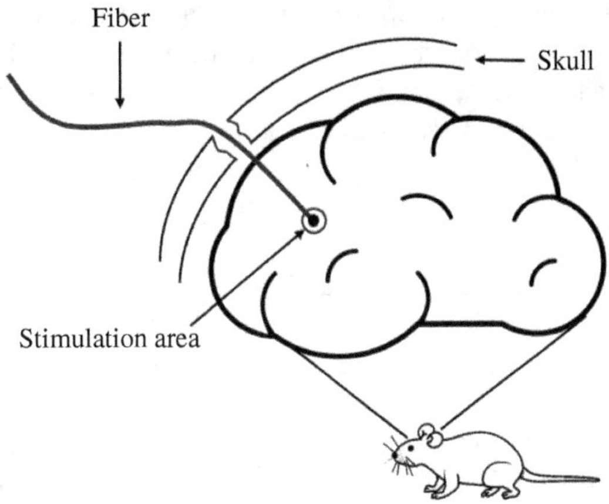

Figure 5.2 Schematic diagram of Laser-generated ultrasound neural regulation via implanted optical fiber.

The current mainstream technical approach for LGUS neural stimulation is centered on an "implantable fiber-coupled transducer" configuration. Its core advantage lies in achieving ultrahigh spatial resolution, thereby enabling precise targeted stimulation at the single-neuron level, as illustrated in Figure 5.2. Benefiting from the ongoing miniaturization of optical fibers, the diameters of associated devices can be reduced to the micrometer scale. Furthermore, the characteristic rapid attenuation of LGUS intensity with propagation distance, combined with the ultra-miniature size of these devices, creates a synergistic effect that jointly ensures the ultrasonic field maintains ultrahigh spatial precision [6, 7]. Existing research has confirmed that the spatial precision of such systems generally reaches the submillimeter level; moreover, by optimizing the geometric configuration of the optical fiber, targeted modulation of subcellular structures such as axons and dendrites can also be achieved. This ultra-microscale manipulation capability is significantly superior to that of traditional piezoelectric ultrasound, which is constrained by physical size limitations and struggles to accomplish targeted stimulation of individual neurons.

The synergistic advantages of the implantable fiber-coupled optical approach are manifested in three key aspects: first, the submillimeter-level sound field enables precise localization of the stimulation region, down to specific individual cells, thereby providing robust support for accurate assessment of stimulation outcomes. Second, the minimally invasive nature significantly reduces surgical trauma and mitigates the risk of associated complications. Third, the optical fiber-based light guide supports remote control and demonstrates excellent compatibility with imaging equipment such as magnetic resonance imaging (MRI). This compatibility enables synchronous monitoring of brain region activation, opening new pathways for in vivo neural modulation research. Together, these advantages underscore the irreplaceability of LGUS compared to electrically driven piezoelectric methods.

Research on neural stimulation using optical fiber-based LGUS technology was attempted in 2020 [6]. A miniaturized fiber-optoacoustic converter (FOC) was developed, whose schematic principle is illustrated in Figure 5.3(a). With a diameter of only 600 μm and a double-layer coating consisting of a ZnO nanoparticle/epoxy diffusion layer and a graphite powder/epoxy absorption layer (Figure 5.3(b)), this design not only achieves omnidirectional laser scattering and efficient photothermal conversion to generate low-frequency ultrasound in the 1–5 MHz range (more suitable for neural modulation) but also overcomes the bulkiness and integration limitations of traditional piezoelectric ultrasound transducers. The FOC is compatible with various experimental modalities such as calcium imaging and electrophysiological recording and is applicable even in implantable and wearable scenarios.

In terms of spatial precision, the ultrasound intensity of the FOC attenuates rapidly with distance, activating neurons only within a 500-μm radius from the fiber tip (Figures 5.3(c–d)). Compared to the millimeter-scale stimulation range of conventional transcranial ultrasound, this represents an order-of-magnitude improvement. The system has successfully accomplished precise mapping of forelimb movement representations in the mouse brain (Figure 5.3(e)), addressing a technical gap in region-specific neuromodulation for small rodents. Furthermore, the technology effectively rules out indirect auditory pathway interference through multiple lines of evidence: in vitro neuron activation experiments performed in the absence of auditory circuits (Figure 5.3(f)), and the lack of response in the contralateral auditory cortex (A1) during in vivo stimulation of the somatosensory cortex (S1) (Figures 5.3(g–h)). Additionally, the laser leakage rate is only 0.11%, and the maximum surface temperature rise is 1.6 °C—well below the thermal activation threshold—confirming that neural activation is directly mediated by ultrasound and avoiding risks associated with nonspecific stimulation.

The FOC also demonstrates excellent biocompatibility and operational stability: it can repeatedly stimulate neurons in vitro for over 2 hours without causing morphological damage, and in vivo stimulation of the motor cortex (M1) induces stable electromyographic responses without tissue injury. Its low duty cycle (0.36%) and safe ultrasonic pressure (0.48 MPa) further support its suitability for long-term behavioral studies in freely

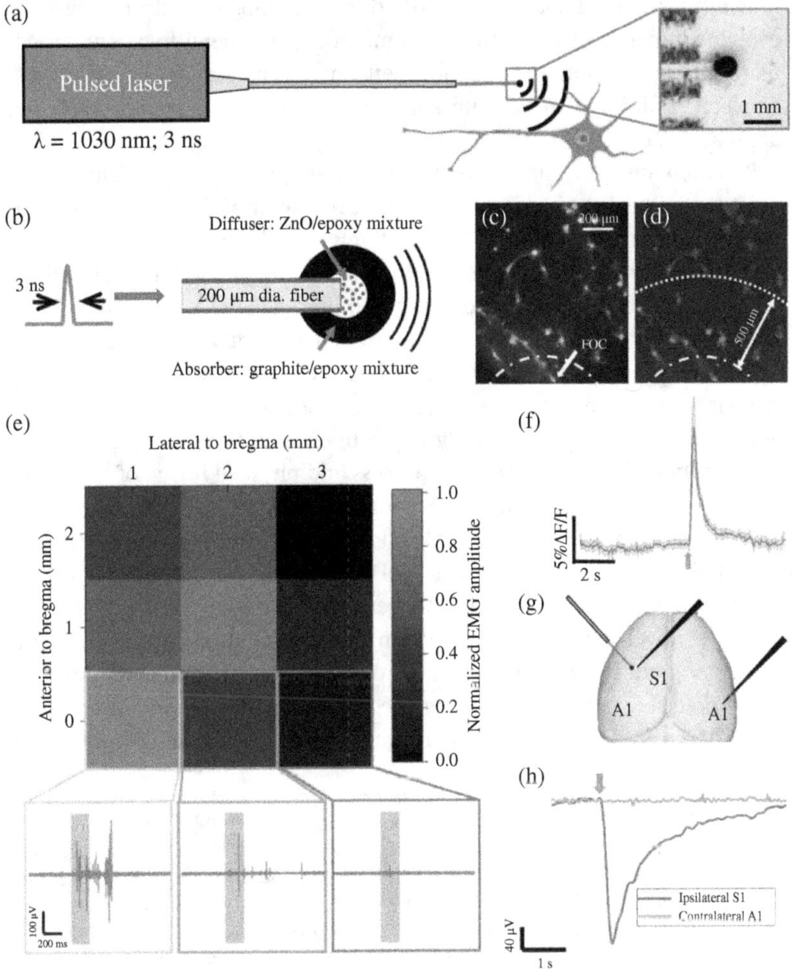

Figure 5.3 FOC for high-precision optoacoustic stimulation [6]. (a) Schematic diagram of the principle of LGUS neuromodulation via FOC. Inset: Magnified FOC tip under a stereomicroscope. (b) Device schematic. (c) and (d) Spatial distribution of the maximum neuronal calcium response induced by FOC stimulation. Dashed line: FOC placement position. (e) Heat map showing the maximum peak-to-peak EMG response amplitude induced by FOC stimulation at different motor cortex positions. Inset: Representative EMG traces at the indicated positions. (f) Average trajectory of calcium signals in neurons in response to 200 ms FOC stimulation (n = 12). Shaded area: ± standard deviation. (g) Place the FOC in S1, and place recording electrodes in the ipsilateral S1 and contralateral A1 to detect the participation of the auditory pathway. (h) Local field potential responses in the ipsilateral S1 region and contralateral A1 region elicited by FOC stimulation in the S1 region. Error bars: ± standard deviation. Arrows: stimulation onset points [6] /

moving animals. Looking forward, this technology holds promise for evolving into a precise, genetically unmodified, and metal-free neural modulation tool for human applications, effectively bridging the technical gap between traditional ultrasound stimulation and invasive electrode-based approaches.

Research on FOC has demonstrated that the spatial precision of its acoustic field is largely determined by the diameter of the fiber tip: smaller tip diameters result in a correspondingly smaller acoustic field range. To further enhance spatial precision, the fiber tip structure was refined and a tapered fiber-optic photoacoustic emitter (TFOE) was developed—an advancement that enables LGUS-mediated activation of subcellular structures, including individual neurons, axons, and dendrites [8]. By optimizing the fabrication process, researchers successfully deposited a uniformly controlled coating approximately 10 μm thick onto the tip of a 20-μm diameter optical fiber. Its fabrication process and physical image of the device are shown in Figure 5.4(a). The reduction in device diameter markedly narrows the acoustic field, not only allowing the simultaneous application of LGUS stimulation and whole-cell patch-clamp recording—which is unachievable with conventional ultrasound technologies, but also providing a novel experimental platform for investigating the mechanisms underlying ultrasound-mediated neural modulation (Figures 5.4(b–c)).

By precisely stimulating target regions containing only axons and dendrites using TFOE, it was confirmed that neurites can be effectively

Figure 5.4 TFOE for high-precision optoacoustic stimulation [8] / Springer Nature / CC BY 4.0. (a) Schematic diagram of the simplified fabrication process for TFOE. Bottom left: Optical image of TFOE; Bottom right: Enlarged view of the tip details. (b–c) Two-photon imaging of GAD2-tdTomato-negative pyramidal neurons and GAD2-tdTomato-positive inhibitory interneurons in mouse brain slices using TFOE-integrated patch-clamp technique. Visualization of patch pipette was achieved through the teal fluorescent dye Alexa Fluor 488 loaded in the intracellular electrode solution. (d–e) TFOE photostimulation triggers synaptic activation accompanied by calcium wave propagation along neural networks. Arrows indicate the target area (rightmost arrow), neuron 1 (bottommost arrow), and neurons 2 and 3 (the remaining two arrows). (f) Calcium signal ΔF/F values of calcium signals at four seconds after laser onset. (g) Calcium signal traces of the target area (rightmost arrow), neuron 1 (bottommost arrow), and neurons 2 and 3 (the remaining two arrows), corresponding to the labels in Figure d. Inset: Magnified view of the transient calcium signal after laser activation. Arrow: Laser activation time. (h) Multipolar neurons receiving selective TFOE stimulation targeting axons (leftmost arrow) and dendrites (remaining two arrows). (i–k) Maximum ΔF values of calcium signals after stimulation in different regions. Scale bars in d–f and h–k: 50 μm.

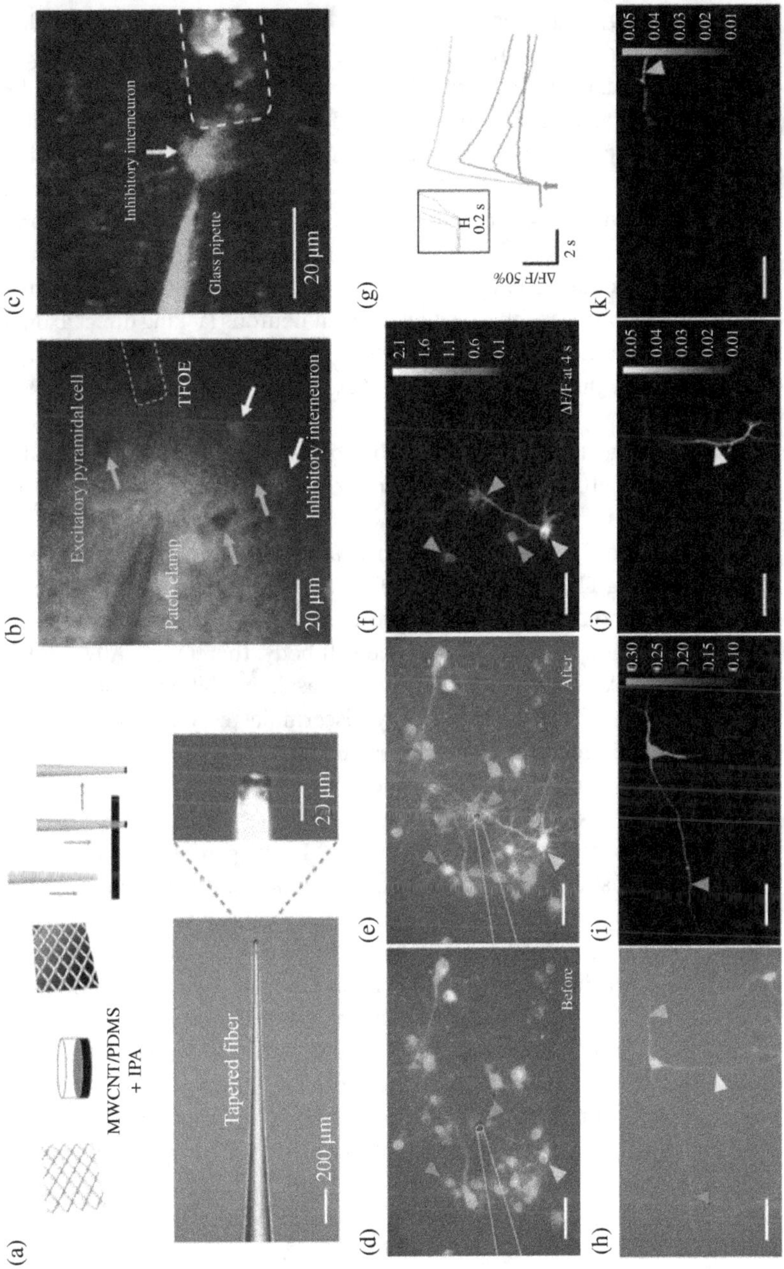
(a)
MWCNT/PDMS
+ IPA
Tapered fiber
200 μm
20 μm
(b)
Excitatory pyramidal cell
Patch clamp
TFOE
Inhibitory interneuron
20 μm
(c)
Inhibitory interneuron
Glass pipette
20 μm
(d)
Before
(e)
After
(f)
2.1
1.6
1.1
0.6
0.1
ΔF/F at 4 s
(g)
0.2 s
ΔF/F 50%
2 s
(h)
(i)
0.30
0.25
0.20
0.15
0.10
(j)
0.05
0.04
0.03
0.02
0.01
(k)
0.05
0.04
0.03
0.02
0.01

activated and recorded three distinct patterns of calcium responses (Figures 5.4(d–g)). First, slow-propagating calcium waves were observed to originate from the target neurons. Second, rapid calcium transients were detected immediately upon stimulation in targeted neurites (marked by right-pointing arrows in Figures 5.4(d–f)) as well as in neurons directly connected to their axons (marked by downward-pointing arrows in Figures 5.4(d–f)). This observation suggests action potential backpropagation—consistent with the finding that no significant temporal difference was observed between the instantaneous calcium signals in the upper and lower traces in Figure 5.4(g). Third, delayed responses of approximately 0.2 seconds were observed in adjacent neurons lacking direct axonal connections (marked by the two left-pointing arrows in Figures 5.4(d–f)), as illustrated in the inset of Figure 5.4(g). These responses are presumed to arise from action potentials induced by synaptic transmission. These findings indicate that TFOE is capable of selectively activating specific subcellular structures and eliciting responses across different levels of neural networks.

Selective stimulation of three neurites of a multipolar neuron was performed using TFOE (Figures 5.4(h)). It was observed that stimulation of the neurite marked by the leftmost arrow elicited a robust and immediate calcium activation response within the cell body, thereby identifying this process as the axon (Figures 5.4(i)). By contrast, stimulation of the remaining two neurites failed to induce any discernible response in the soma, confirming their identity as dendrites (Figures 5.4(j–k)). By harnessing the subcellular spatial precision afforded by TFOE, the researchers were able to resolve distinct dynamic response profiles of axons versus dendrites to LGUS stimulation. Moreover, the TFOE platform is capable of generating acoustic pulses on a sub-microsecond timescale from a single, 3-nanosecond optical laser pulse, representing a substantial advance in the temporal resolution achievable with acoustic neuromodulation technology.

In addition, the integration of LGUS stimulation and electrophysiological recording enables a bidirectional brain–computer interface (BCI), capable of "writing" through neural stimulation and "reading" via neural activity recording. This led to the development of a multifunctional fiber-based optoacoustic emitter (mFOE) [9] As shown in Figure 5.5, the mFOE operates by converting laser pulses into acoustic waves, while its embedded electrodes simultaneously capture neural activity and transduce it into electrical signals. A key feature of this system is the orthogonality between the emitted sound waves and the associated electric fields, which effectively eliminates the signal interference commonly observed between

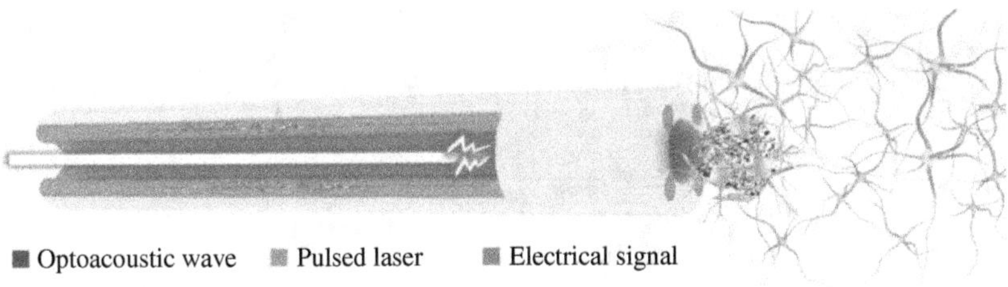

Figure 5.5 Schematic diagram of the mFOE structure [9] / with permission of John Wiley & Sons.

electrical stimulation and recording modalities. This development opens new avenues for neural circuit investigation and advancing research in ultrasonic neuromodulation.

While implantable fiber-optic light-guiding systems have enabled near-field modulation with single-neuron precision, their invasiveness and the requirement for craniotomy continue to limit clinical translation. Meanwhile, noninvasive transcranial focused ultrasound (tFUS) still faces fundamental technical challenges in achieving sufficient spatial resolution. To address these limitations, a soft optoacoustic pad (SOAP) was developed. Through careful geometric design, the SOAP achieves an ultrahigh numerical aperture of 0.95—approaching the theoretical limit of 1—as illustrated in Figure 5.6(a). This design enables the generation of 15 MHz single-cycle-focused ultrasound that, even after transcranial transmission, maintains a lateral resolution of 83 μm and an axial resolution of 287 μm. These values represent a two-order-of-magnitude improvement over the millimeter-scale resolution typical of conventional tFUS. The focal region measures only 100 μm in vitro and 200 μm in vivo, allowing precise targeting of small neuronal populations [10].

A key feature of the SOAP is its single-cycle ultrasound waveform, which eliminates the need for multi-cycle accumulation. This not only improves spatiotemporal precision but also reduces the acoustic energy density required for neural stimulation to as low as 5.7×10^{-4} J/cm^2—four orders of magnitude lower than that of conventional tFUS. Moreover, the system achieves a transcranial efficiency of 69%, with no significant degradation in focusing precision after passing through the mouse skull. Importantly, it operates without surgical implantation or cranial windows. In terms of neural modulation efficacy, in vitro experiments demonstrated a 71.4% activation success rate in cultured cortical neurons, eliciting two distinct response patterns: transient (decay time 4.2 s) and sustained (decay time 7.8 s). Activation was strictly confined to the focal spot, and

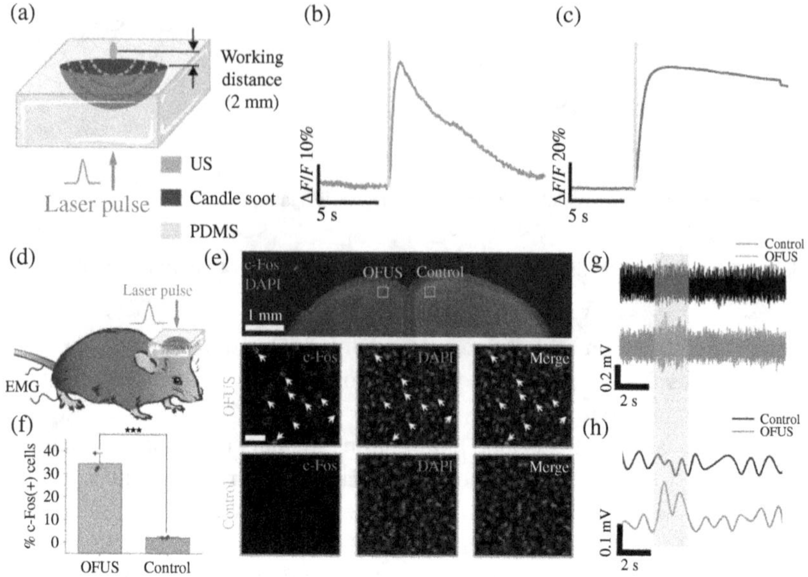

Figure 5.6 Soft optoacoustic pad [10]. (a) Schematic diagram of SOAP. (b) Average calcium ion signal trajectory of transient stimulation. (c) Average calcium ion signal trajectory of continuous focused LGUS stimulation. The solid line represents the average trajectory, and the shaded area represents the standard error. Vertical line: laser on. (d) Schematic diagram of in vivo test stimulation. (e) Representative images of c-Fos and DAPI staining in the stimulation area and the control area. DAPI is used to label all neurons, while c-Fos is used to label activated neurons. The arrows in the figure indicate the neurons that have been activated and stained via focused LGUS. Left outline: OFUS stimulation area. Right outline: contralateral area control group. Scale bar: middle and lower images, 50 μm. (f) Statistical analysis of the percentage of c-Fos positive neurons. ***$p<0.001$, two-sample t-test. (g) Representative EMG recordings of 2-second OFUS stimulation on the somatosensory cortex and the control group. Rectangle: laser on. (h) EMG signals after band-pass filter, full-wave rectification and envelope processing [10] / Springer Nature / CC BY 4.0.

the corresponding calcium signals are shown in Figure 5.6(b) and (c), respectively. In vivo experiments (setup shown in Figure 5.6(d)), validated by c-Fos immunofluorescence staining (Figure 5.6(e)), revealed a stimulation region diameter of only 200 μm in the mouse motor cortex, with 34% of neurons being c-Fos positive compared to only 2% in the control group (Figure 5.6(f)). Electromyography (EMG) recordings further confirmed that the stimulation reliably evoked contralateral hindlimb movements (Figure 5.6(g–h)), with no observed off-target activation or auditory artifacts. Safety assessments were conducted at both physical and biological levels. Physically, the mechanical index (0.5) remained well below the FDA limit, the focal temperature rise was less than 0.1 K, and no cavitation risk was detected. Biologically, neuron viability was unaffected by repeated stimulation in vitro, and H&E staining of brain tissue poststimulation showed no morphological damage, confirming the safety profile required for noninvasive applications.

Together, the SOAP platform provides a new tool for neural circuit analysis and the treatment of neurological disorders, offering a unique combination of ultrahigh precision, noninvasiveness, and demonstrated safety.

Although SOAP has achieved noninvasive transcranial LGUS-based neural stimulation, skull-induced aberration remains a major obstacle to its broader adoption and application. This challenge is particularly pronounced in humans due to the relatively thick skull, which tends to cause significant acoustic distortion that must be compensated via acoustic wavefront adjustment. The core issue stems from the acoustic attenuation and phase distortion introduced by the skull medium, necessitating the development of precise phase-correction algorithms and dedicated hardware to enable efficient transcranial acoustic energy delivery. The wavefront shaping technology based on LGUS offers the theoretical potential for sub-wavelength precision focusing, presenting distinct advantages. Figure 5.7 illustrates a schematic of transcranially focused LGUS neural stimulation implemented through acoustic field wavefront manipulation.

The core challenge faced by tFUS stems from the high acoustic attenuation coefficient and complex acoustic heterogeneity of the skull. This leads to phase distortion and energy dissipation during sound wave propagation, which in turn causes focal point shift and focal spot splitting. Meanwhile,

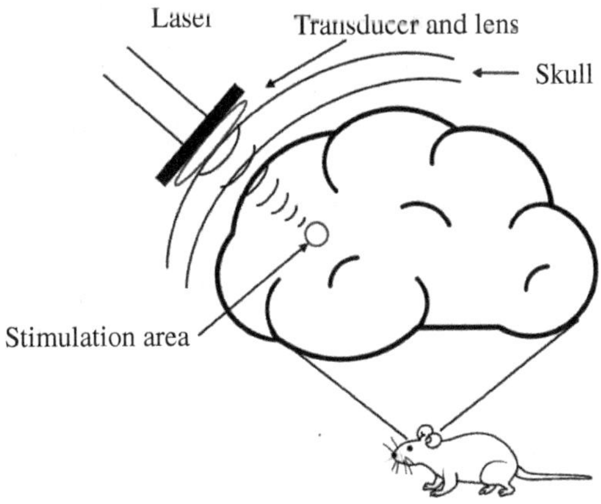

Figure 5.7 Schematic diagram of noninvasive transcranial focused laser-generated ultrasound neural regulation.

the longitudinal wave-transverse wave coupling (multimodal waveform conversion) during propagation changes the position of the focal point, further increasing the difficulty of focusing. Traditional solutions rely on craniotomy to remove the interference of the skull, but this violates the principle of noninvasive treatment. In recent years, with the development of multimodal measurement technologies such as nuclear magnetic resonance and CT, combined with sound velocity distribution modeling, it has become possible to accurately reconstruct the three-dimensional acoustic parameter map of the skull, providing high-fidelity boundary conditions for sound field simulation.

The current mainstream simulation algorithm for transcranial focusing is based on time-reversal technology, which exploits the time reversibility of acoustic wave propagation. Its core procedure consists of three stages: first, constructing a physical model using skull acoustic data acquired via CT imaging; second, placing a virtual point source at the intended focal target and positioning virtual sensors at the transducer to capture the resulting acoustic signals; third, time-reversing the recorded signals and retransmitting them to achieve acoustic energy focusing at the target point (as illustrated in Figure 5.8: panel a depicts acquisition of transcranial propagation data from the virtual point source, and panel b illustrates the reverse emission to accomplish focusing). Based on the phase shifts derived from this method, an acoustic lens can be designed to enable precise focusing control. While the principle of time reversal is straightforward,

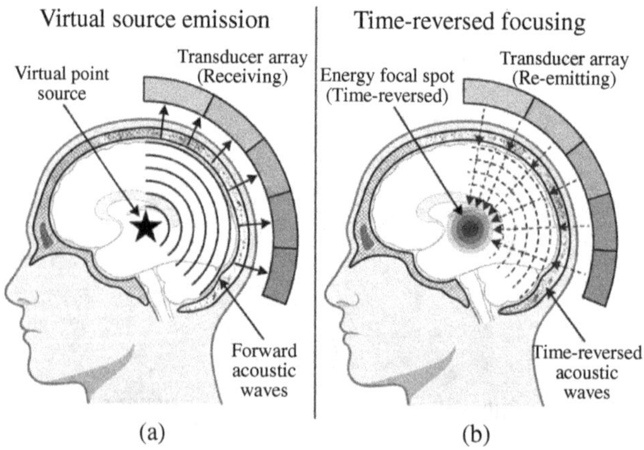

Figure 5.8 Schematic diagram of the time-reversal technique. (a) Acquire transcranial ultrasound propagation data of virtual point sound sources. (b) Output ultrasound in reverse to achieve focusing.

its practical implementation faces two central challenges: first, individual skull variations—including morphology, density distribution, and other acoustic-specific properties—alter sound wave propagation paths, causing discrepancies between simulated models and actual biological tissue responses; second, nonideal deviations between the transducer's emitted waveform and the theoretical waveform—such as phase jitter and amplitude fluctuation—degrade focusing precision. Although these factors do not completely disrupt focusing, they lead to issues like focal spot displacement and energy dispersion. Therefore, how to integrate waveform changes into time reversal method simulations, and how to promote specific transcranial focusing to a universal scheme (i.e., fine-tuning the focusing strategy according to skull parameters) are important research directions in the future.

The technical solutions described above have demonstrated the high-precision advantages of LGUS neuromodulation, yet the specific underlying mechanisms remain incompletely understood. Current hypotheses regarding LGUS-induced neuronal activation include mild thermal effects, radiation force, and cavitation. Among these, membrane heating due to ultrasound-induced temperature rise was once considered a central mechanism. However, subsequent studies have shown that the temperature increase during LGUS stimulation is negligible—far below the conventional thermal threshold required for neural excitation—thus largely ruling out thermal effects as the primary driver. Cavitation-based mechanisms propose that ultrasound disturbs neuronal membrane structure, leading to capacitive currents. Previous studies suggest that continuous ultrasound can induce intracellular cavitation, altering membrane capacitance and generating current via a "bilayer sonoporation model" grounded in bubble dynamics and cellular biomechanics [11, 12]. However, stable cavitation typically requires 2–12 acoustic cycles to establish consistent oscillation. Notably, it has been demonstrated that single-cycle-focused LGUS can still elicit neural response [10]. The single-cycle stimulation time is too short to form stable bubbles and maintain oscillation to interfere with ion channels. Therefore, the cavitation effect induced by continuous cycles is not the only mechanism of action.

Currently, the mainstream view holds that the activation of mechanically sensitive ion channels is the core mechanism of LGUS neuromodulation [13] (see schematic in Figure 5.9). Mechanosensitive ion channels are membrane proteins that respond to mechanical stimuli: when ultrasound acts on neurons, it induces mechanical vibration of the cell membrane, altering its physical properties and thereby activating these channels. Once opened,

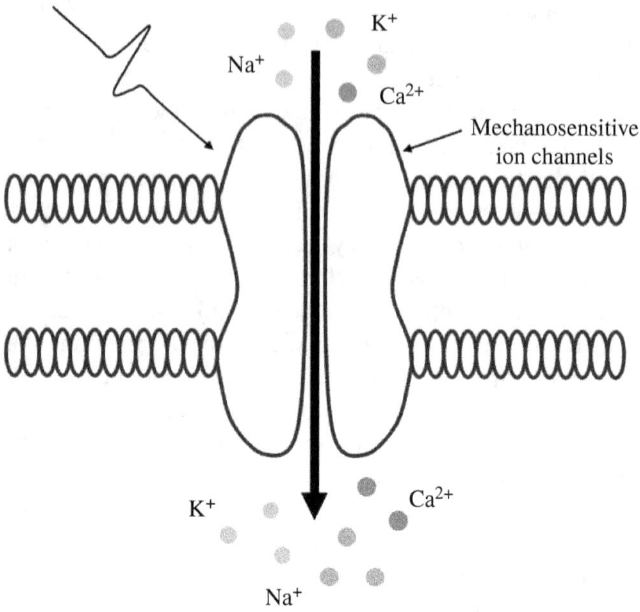

Figure 5.9 Laser-generated ultrasound stimulation of mechanosensitive ion channels for neural activation.

ion currents accumulate, leading to membrane depolarization, subsequent activation of voltage-gated sodium channels, and ultimately the initiation of action potentials. Experimental evidence further supports this model: the energy required for LGUS stimulation is only 1/40 of that needed for photothermal excitation. Coupled with analysis of LGUS device energy-conversion efficiency, this further excludes significant contributions from thermal effects.

Key evidence supporting this mechanism stems from the analysis of action potential characteristics. While patch-clamp technology serves as a standard approach for directly recording subthreshold and suprathreshold neuronal electrical activity, the limited focusing precision of traditional ultrasound often disrupts the seal between electrode and cell membrane, making intracellular patch-clamp recording difficult. In contrast, LGUS—benefiting from its high spatial resolution—has achieved compatibility with whole-cell patch-clamp configurations, enabling detection of single-neuron electrophysiological responses. Experimental data reveal that action potential waveforms elicited by LGUS stimulation closely resemble those of spontaneous neuronal activity—a resemblance not observed under photothermal stimulation, suggesting that the two may share similar biophysical mechanisms. Further verification comes from experiments targeting specific ion-channel regulation. When mechanosensitive ion channels were blocked or overexpressed, LGUS-induced electrical responses changed significantly. This dose-dependent effect provides direct experimental support for the central role of mechanosensitive ion channels, while excluding confounding factors such as nonspecific thermal contributions.

In addition to modulating the central nervous system, LGUS neurostimulation technology can also be extended to high-precision intervention of peripheral sensory organs. For instance, in the field of visual system restoration, a flexible LGUS device was applied for retinal neural stimulation [14]. Such device employs a composite film with PDMS as the matrix and carbon materials (such as candle soot or carbon nanotubes) as the light-absorbing layer. When irradiated by near-infrared pulsed lasers, the film generates a localized ultrasonic field, thus precisely stimulating the mechanosensitive neurons in the retina. It has been demonstrated that the lateral resolution of the generated ultrasonic field can reach 51 μm, which can effectively modulate the activity of retinal ganglion cells both in vitro and in vivo, and even elicit significant responses in retinal degenerative disease models. In living animals, the implantation of the film under the retina triggered neural signals in the superior colliculus, confirming its potential to effectively drive the visual pathway. This strategy boasts multiple advantages: the acoustic pressure threshold required for stimulation is as low as 0.05 MPa, and the temperature rise induced is less than 1 °C, which complies with ophthalmic safety standards. Its coupling mechanism of optical excitation and acoustic conduction combines the spatial continuity of optical modulation with the deep-tissue penetration capability of ultrasound, laying a foundation for the realization of high-density, high-resolution neural interfaces. Moreover, the flexible design of the device enhances its biocompatibility and implantation stability. It not only opens up a new avenue for the development of retinal prostheses but also provides a scalable technical platform for other neuromodulation scenarios—such as peripheral nerve interfaces, and closed-loop neural repair systems—demonstrating the broad application prospects of LGUS neuromodulation in next-generation bioelectronic medicine.

5.2 Drug Delivery

In biomedical applications, drug delivery systems must meet the core requirements of being noninvasive, precise, and spatiotemporally controllable. This entails avoiding tissue damage from invasive procedures, ensuring on-demand drug release at target sites, and minimizing adverse effects on healthy tissues. LGUS technology fulfills these needs with key advantages such as tightly focused acoustic spots, broadband frequency response, and negligible ultrasound-induced thermal effects. Within the field of drug delivery, LGUS primarily advances two critical objectives: first, the precise remote manipulation of microparticles and liquids, addressing the challenge of accurate drug carrier transport to target areas; and second, the spatiotemporally controlled release from nanoscale drug delivery systems, enabling on-demand therapeutic activation. Together, these capabilities offer a promising new direction for innovation in targeted drug delivery technologies.

Precise directional and long-range manipulation of microscopic particles forms the foundation for targeted drug delivery. Only when carrier particles accurately reach the lesion site can subsequent efficient drug release be achieved. Among current mainstream particle manipulation technologies, light-driven methods enable noncontact control, yet they are often constrained by contact-line pinning effects [15], which can hinder the movement of liquids and particles and impede long-distance transport. In contrast, ultrasound technology has been widely applied in microfluidic systems. To generate sufficient acoustic pressure for driving microparticles under non-focused conditions, improving the energy conversion efficiency from light to ultrasound is essential. An innovative technical approach has been developed to achieve high-efficiency LGUS conversion, significantly enhancing acoustic pressure and enabling diverse forms of microparticle manipulation without the need for focusing [16]. The feasibility of this method is demonstrated experimentally. Schematic diagrams of the setup are shown in Figure 5.10(a–b), with corresponding optical images in Figure 5.10(c–h). The experiment employs a transparent microfluidic channel loaded with a group of microparticles. When the laser is off from 0 to 4 seconds, the particles remain stationary (Figure 5.10(c)). At 4 seconds, the laser is activated, and the LGUS transducer immediately generates ultrasound, initiating directional motion of the microparticle group. Subsequent observations at 8, 10, 12, and 14 seconds show the particle swarm continuing to advance along the direction of laser irradiation, demonstrating stable, long-range directional manipulation (Figure 5.10(d–h)). Control experiments further confirmed that particle motion is solely attributable to the mechanical effects of ultrasound, with no direct correlation to the laser itself—thus eliminating potential confounding factors such as photothermal or photochemical effects. This technology not only overcomes the movement limitations associated with light-driven

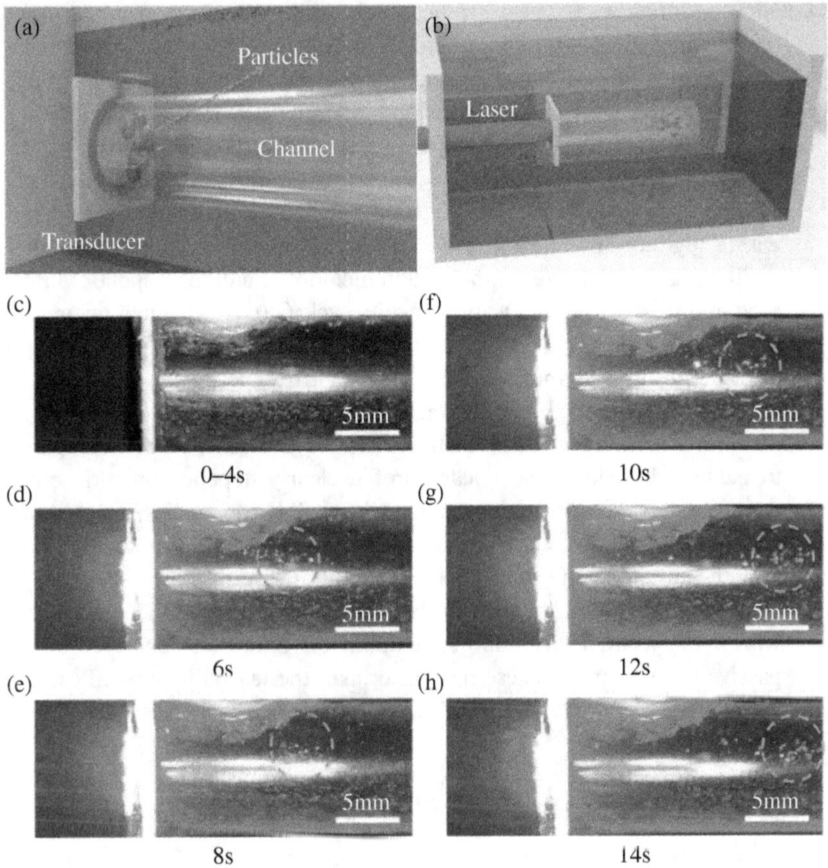

Figure 5.10 (a) Schematic diagram of the photoacoustic control system. (b) Schematic diagram of particle motion under LGUS wave action. (c–h) Optical images of particle manipulation processes at different time points [16] / with permission of Elsevier.

approaches but also achieves non-contact, directionally controllable, and long-distance particle manipulation, offering reliable technical support for precise drug carrier delivery.

After drug carriers successfully reach the target area, on-demand release becomes the next critical requirement. Achieving precise drug release at predetermined times and locations through external control can significantly enhance therapeutic efficacy while minimizing side effects [17]. Among remotely controllable drug release technologies, ultrasound-triggered systems offer several advantages: they are operationally portable, allow modulation of release rate by adjusting acoustic power and duty cycle, and can prevent premature drug leakage. These attributes make them particularly suitable for localized delivery of proteins, polypeptides, and small-molecule drugs. A spatiotemporally controlled drug delivery system based on LGUS employs a laser-activated carbon black/PDMS device to generate high-frequency, high-pressure pulsed ultrasound. This triggers drug release from drug-loaded PLGA nanoparticles encapsulated within alginate microgels through combined cavitation effects and shell oscillations. Key features of the system include a small focal spot (<1 mm), high frequency (>10 MHz), and low duty cycle (<0.001%), enabling precise remote spatiotemporal control of drug release while substantially reducing ultrasound-induced thermal damage. In vitro experiments demonstrate that this system significantly increases both the rate and cumulative amount of drug release, achieving over threefold higher release compared to passive diffusion. The release profile clearly depends on ultrasound energy and exposure duration. In biological validation studies, the medium released under LGUS activation showed stronger anti-proliferative and growth-inhibitory effects on HeLa cells and derived tumor spheroids, as well as significantly enhanced antibacterial activity against *Escherichia coli*, with notably larger inhibition zone diameters than those observed in the passive release group. These results confirm the feasibility and efficacy of the LGUS-based system for localized, on-demand drug delivery.

5.3 Tissue Cutting and Ablation

By enhancing the acoustic pressure of LGUS, high-precision tissue-targeted therapy can be achieved through controlled cavitation effects. The cavitation process exerts mechanical forces capable of splitting cells or breaking down tissues into cell-free homogenates without causing damage to surrounding healthy structures. Compared with the thermal ablation mechanism of traditional high-intensity-focused ultrasound (HIFU), this approach significantly improves both treatment accuracy and safety. While pulsed-laser-induced cavitation relies on optical penetration depth and is thus limited to transparent or superficial tissues, the acoustic cavitation generated by LGUS is not constrained by tissue transparency and can effectively reach deeper anatomical regions. The core mechanism involves differential effects within the ultrasound focal zone. Cells located at the center of the focal region are precisely disrupted under the combined action of focused shockwaves and cavitation-induced microstreaming, whereas peripheral cells are radially dispersed by cavitation-driven fluid jets. This enables directional tissue manipulation with spatial selectivity [18]. This section focuses on three key application scenarios of LGUS-mediated cavitation: precise tissue cutting, mechanical tissue ablation, and thrombolysis. Each will be systematically elaborated in terms of technical implementation, experimental validation, and clinical or translational value.

High-precision micro-cutting represents a fundamental requirement in microsurgery for treating diseased tissues, such as during brain tumor resection. Traditional laser cutting, which relies on dielectric breakdown, exhibits limited efficacy in opaque or deep-seated tissues. Early micro-cavitation techniques also necessitated placing target tissues on rigid acoustic reflectors to amplify ultrasonic intensity, rendering them unsuitable for deep internal tissues distant from rigid substrates. LGUS overcomes this bottleneck through innovative device design: by coating a laser-to-ultrasound conversion material onto the concave surface of a plano-concave optical lens, it generates a focused LGUS pulse with a center frequency f_c = 15 MHz and a peak negative pressure amplitude $P_- > 30$ MPa at the focal point. This pulse intensity is sufficient to induce free-field microcavitation in aqueous media without requiring a rigid substrate [19]. This substrate-independent cavitation capability allows LGUS to act directly on deep internal tissues, enabling high-precision cutting. For instance, in targeted brain surgery, tumor tissues can be mechanically isolated from surrounding healthy tissue via cavitation effects. Subsequent loss of blood supply and nutrients leads to gradual necrosis of the diseased tissue, avoiding the neural trauma often associated with conventional surgical resection. The core advantages of this approach are precision and controllability: cavitation effects are confined to the ultrasonic focal zone, and when combined with the submillimeter focusing capability of LGUS, enable "point-to-point" ablation of minute pathological structures—all without thermal damage risks. This positions LGUS as a promising new tool for minimally invasive microsurgery [19, 20].

Beyond the precise excision of localized lesions, the cavitation effect produced by LGUS can also be applied to achieve mechanical ablation of entire tissue volumes. By utilizing cavitation-induced mechanical forces, LGUS can thoroughly decompose target tissues into acellular homogenates, making it particularly suitable for clinical scenarios requiring complete removal of diseased tissue as illustrated schematically in Figure 5.11. Experimental studies have validated the feasibility of this approach. In ablation tests performed on intestinal cell spheroids and porcine ocular

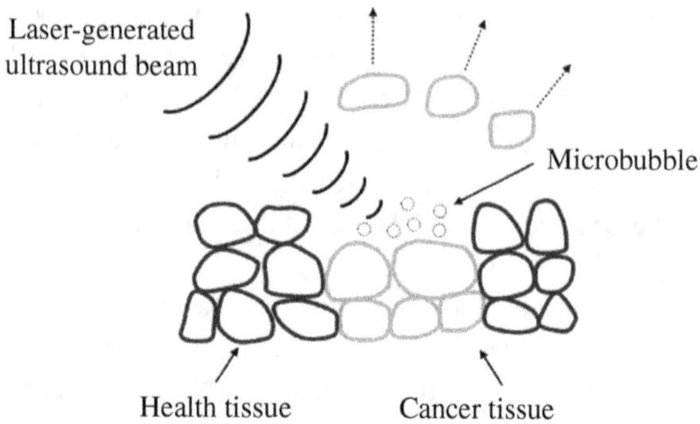

Figure 5.11 Illustration of the cutting of cancerous tissue using focused laser-generated ultrasound beams combined with microbubbles.

tissue, LGUS successfully achieved uniform tissue decomposition via sustained cavitation activity without damaging the surrounding healthy structures [19]. However, current implementations face distinct limitations: the ablation zone per treatment cycle remains relatively small, and an inter-treatment spacing of approximately 50 μm is required to achieve optimal outcomes. Consequently, the time needed for ablating larger tissue volumes becomes considerably extended, which constrains clinical practicality and efficiency. To address this challenge, a clear research direction has been identified: employing lasers with higher pulse repetition frequencies can reduce the required accumulation time for cavitation effects, thereby enhancing the efficiency of large-scale tissue ablation. This strategy provides a promising pathway for optimizing the technology and advancing its clinical translation.

In addition to the previously described capabilities of cutting or ablating tissue cells, thrombolysis represents another significant application extended from the cavitation effect of LGUS [21]. Compared with traditional pharmacological therapies or thrombectomy, ultrasonic thrombolysis offers advantages such as high thrombolytic efficiency and favorable safety profiles. Furthermore, the synergistic interaction with ultrasound-activated microbubbles can further shorten the duration of thrombolysis and reduce the required ultrasound intensity. Leveraging its features of miniaturization and high acoustic pressure output, LGUS has led to the development of two core technical approaches in the field of thrombus dissolution: LGUS patch-mediated thrombolysis and dual-frequency ultrasound synergistic thrombolysis.

A self-healing flexible LGUS patch, fabricated using carbon nanotubes and PDMS, is capable of generating high-intensity ultrasound exceeding 25 MPa without the need for a focusing structure, thereby delivering stable acoustic energy output for thrombolysis experiments [22]. In the experimental setup, a mirror was employed to direct the laser onto the LGUS patch attached to a transparent petri dish, with the thrombus sample positioned above the patch. A schematic diagram of the experimental design and device is presented in Figure 5.12(a–b). To evaluate thrombolytic efficacy under varying conditions, two thrombus models—thin and thick—were utilized. For the thin thrombus model, as depicted in Figure 5.12(c–d), within 20 minutes of ultrasound exposure, a notable reduction in clot diameter was observed, accompanied by expansion of surrounding blood staining. The normalized mass decreased continuously over time, with the rate of reduction gradually slowing, as illustrated in Figure 5.12(e). In contrast, for the thick thrombus model (Figure 5.12(f–g)), after 40 minutes of ultrasound application, multiple depressions appeared on the clot surface and its thickness diminished. The time course of normalized mass reduction was longer compared to that of the thin thrombus as shown in Figure 5.12(h). This difference can be attributed to the greater ease with which ultrasound waves penetrate thin thrombi, coupled with their larger contact area with microbubbles, resulting in higher thrombolytic efficiency.

Intravascular thrombolysis imposes strict limitations on device dimensions: conventional piezoelectric ultrasound faces challenges in achieving high frequencies under miniaturized conditions, whereas LGUS can maintain high-frequency output even in miniaturized devices. Leveraging this distinction, a laser-coupled dual-excitation ultrasonic transducer integrating LGUS and piezoelectric ultrasound technologies has been developed. This design can generate enhanced cavitation effects under the same or lower input power, significantly improving thrombus ablation efficiency [23]. The core innovation lies in high-frequency complementarity: LGUS provides high-frequency output, while piezoelectric ultrasound supplies complementary low-frequency acoustic energy. Through synergistic interaction, the number and oscillatory intensity of cavitation bubbles are substantially increased, accelerating thrombus disintegration. A schematic diagram is provided in Figure 5.13. Experimental results demonstrate that, compared to using LGUS or piezoelectric ultrasound alone, the dual-frequency ultrasound technology markedly enhances both mass reduction and dissolution rate of thrombus within the same timeframe. This approach is particularly suited for intravascular thrombolysis applications. The miniaturized device design accommodates intravascular operational constraints, and the high-frequency synergistic effect addresses the insufficient thrombolytic efficiency associated with single-mode technologies, thereby exhibiting considerable potential for clinical application.

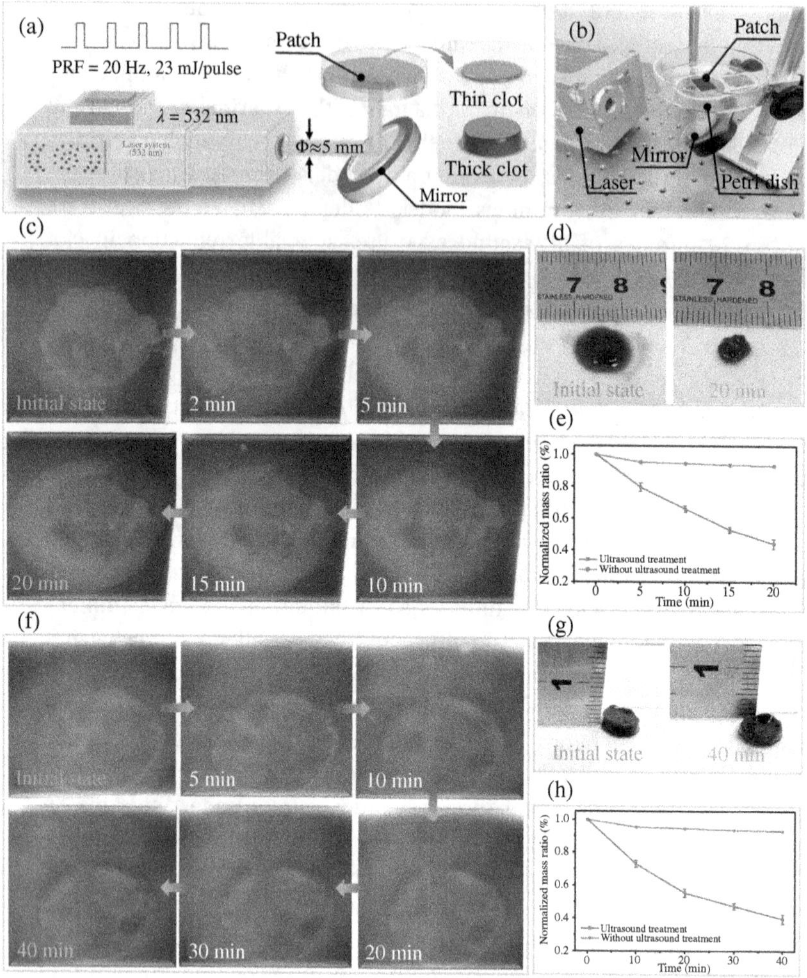

Figure 5.12 Thrombolytic experiment of self-healing LGUS patch [22] / Springer Nature / CC BY 4.0. (a) Schematic diagram of experimental design. (b) Setup of the experiment. (c) State changes of thin blood clots at the initial moment, 2nd, 5th, 10th, 15th, and 20th minutes of LGUS action. (d) Thin blood clot in its initial state and after 20 minutes of ultrasound treatment. (e) Normalized mass change curve of thin blood clots, in which the control group (the upper line) without ultrasound treatment and the experimental group (the lower line) with ultrasound treatment ($n = 4$). (f) Changes in the state of thick blood clots at the initial moment, 5th, 10th, 20th, 30th, and 40th minutes of LGUS application. (g) Thick blood clot in its initial state and after 40 minutes of ultrasound treatment. (h) Normalized mass change curve of thick blood clots, in which the control group (the upper line) without ultrasound treatment and the experimental group (the lower line) with ultrasound treatment ($n = 4$).

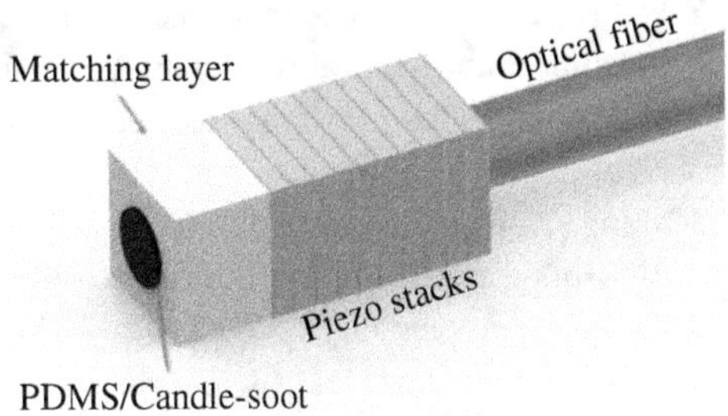

Figure 5.13 Schematic diagram of a laser-coupled dual excitation ultrasonic transducer [23] / with permission of IEEE.

5.4 Summary

This chapter systematically elaborates on the three core applications of LGUS technology in the medical field: neuromodulation, drug delivery, and tissue cutting and ablation. By precisely coupling submillimeter-focused acoustic fields with mechanosensitive ion channels, neuromodulation technology achieves noninvasive stimulation at the single-neuron level, with its high spatiotemporal resolution and compatibility with MRI environments providing revolutionary tools for neuroscientific research. The noninvasive manipulation characteristics of high-frequency LGUS mechanical waves open new pathways for the directional movement of microfluidic particles and the spatiotemporally controlled release of nanoscale drug delivery systems. Meanwhile, high-intensity focused pulses, through the synergistic effects of cavitation and shockwaves, enable deep-tissue cellular-level decomposition without thermal damage, significantly improving treatment precision and safety. Although challenges remain in material engineering optimization, transcranial acoustic field correction, and the analysis of multi-mechanism coupling effects, continuous breakthroughs in broadband response, miniaturized design, and energy conversion efficiency of LGUS are driving the technology from fundamental research toward clinical translation. In the future, interdisciplinary collaborative innovation will accelerate the deep integration of LGUS in multiscale biomechanical regulation, offering novel technological paradigms for precision medicine and intelligent diagnostics.

References

1 Wang, G., Zhou, Y., Yu, C., Yang, Q., Chen, L., Ling, S., and Zhao, Q. (2024). Intravital photoacoustic brain stimulation with high-precision. *Journal of Biomedical Optics*, *29*(S1), S11520–S11520.

2 Pellow, C., Pichardo, S., and Pike, G. B. (2024). A systematic review of preclinical and clinical transcranial ultrasound neuromodulation and opportunities for functional connectomics. *Brain Stimulation*, *17*(4), 734–751.

3 Meng, Y., Hynynen, K., and Lipsman, N. (2021). Applications of focused ultrasound in the brain: from thermoablation to drug delivery. *Nature Reviews Neurology*, *17*(1), 7–22.

4 Sarica, C., Nankoo, J. F., Fomenko, A., Grippe, T. C., Yamamoto, K., Samuel, N., and Chen, R. (2022). Human studies of transcranial ultrasound neuromodulation: a systematic review of effectiveness and safety. *Brain Stimulation*, *15*(3), 737–746.

5 Zhang, T., Wang, Z., Liang, H., Wu, Z., Li, J., Ou-Yang, J., and Zhu, B. (2022). Transcranial focused ultrasound stimulation of periaqueductal gray for analgesia. *IEEE Transactions on Biomedical Engineering*, *69*(10), 3155–3162.

6 Jiang, Y., Lee, H. J., Lan, L., Tseng, H. A., Yang, C., Man, H. Y., and Cheng, J. X. (2020). Optoacoustic brain stimulation at submillimeter spatial precision. *Nature Communications*, *11*(1), 881.

7 Chen, G., Shi, L., Lan, L., Wang, R., Li, Y., Du, Z., and Yang, C. (2022). High-precision neural stimulation by a highly efficient candle soot fiber optoacoustic emitter. *Frontiers in Neuroscience*, *16*, 1005810.

8 Shi, L., Jiang, Y., Fernandez, F. R., Chen, G., Lan, L., Man, H. Y., and Yang, C. (2021). Non-genetic photoacoustic stimulation of single neurons by a tapered fiber optoacoustic emitter. *Light: Science & Applications*, *10*(1), 143.

9 Zheng, N., Jiang, Y., Jiang, S., Kim, J., Chen, G., Li, Y., and Yang, C. (2023). Multifunctional fiber-based optoacoustic emitter as a bidirectional brain interface. *Advanced Healthcare Materials*, *12*(25), 2300430.

10 Li, Y., Jiang, Y., Lan, L., Ge, X., Cheng, R., Zhan, Y., and Cheng, J. X. (2022). Optically-generated focused ultrasound for noninvasive brain stimulation with ultrahigh precision. *Light: Science & Applications*, *11*(1), 321.

11 Krasovitski, B., Frenkel, V., Shoham, S., and Kimmel, E. (2011). Intramembrane cavitation as a unifying mechanism for ultrasound-induced bioeffects. *Proceedings of the National Academy of Sciences*, *108*(8), 3258–3263.

12 Plaksin, M., Shoham, S., and Kimmel, E. (2014). Intramembrane cavitation as a predictive bio-piezoelectric mechanism for ultrasonic brain stimulation. *Physical Review X*, *4*(1), 011004.

13 Chen, G., Yu, F., Shi, L., Marar, C., Du, Z., Jia, D., and Yang, C. (2024). High-precision photoacoustic neural modulation uses a non-thermal mechanism. *Advanced Science, 11*(32), 2403205.

14 Leong, A., Li, Y., Ruikes, T.R., Voillot, J., Yuan, Y., Chen, G., and Picaud, S. (2026) A flexible photoacoustic retinal prosthesis. *Nat Commun* 17, 815, https://doi.org/10.1038/s41467-025-67518-6.

15 Lv, J., Liu, Y., Wei, J., Chen, E., Qin, L., and Yu, Y. (2016). Photocontrol of fluid slugs in liquid crystal polymer microactuators. *Nature, 537*, 179–184.

16 Chen, Z., Wu, Y., Yang, Y., Li, J., Xie, B., Li, X., and Zhu, B. (2018). Multilayered carbon nanotube yarn based optoacoustic transducer with high energy conversion efficiency for ultrasound application. *Nano Energy, 46*, 314–321.

17 Di, J., Kim, J., Hu, Q., Jiang, X., and Gu, Z. (2015). Spatiotemporal drug delivery using laser-generated-focused ultrasound system. *Journal of Controlled Release, 220*, 592–599.

18 Baac, H. W., Lee, T., and Guo, L. J. (2013). Micro-ultrasonic cleaving of cell clusters by laser-generated focused ultrasound and its mechanisms. *Biomedical Optics Express, 4*(8), 1442–1450.

19 Lee, T., Luo, W., Li, Q., Demirci, H., and Guo, L. J. (2017). Laser-induced focused ultrasound for cavitation treatment: toward high-precision invisible sonic scalpel. *Small, 13*(38), 1701555.

20 Baac, H. W., Frampton, J., Ok, J. G., Takayama, S., and Guo, L. J. (2013). Localized micro scale disruption of cells using laser-generated focused ultrasound. *Journal of Biophotonics, 6*(11-12), 905–910.

21 Baac, H. W., Ok, J. G., Maxwell, A., Lee, K. T., Chen, Y. C., Hart, A. J., and Guo, L. J. (2012). Carbon-nanotube optoacoustic lens for focused ultrasound generation and high-precision targeted therapy. *Scientific Reports, 2*(1), 989.

22 Zhang, T., Li, C. H., Li, W., Wang, Z., Gu, Z., Li, J., and Zhu, B. (2024). A self-healing optoacoustic patch with high damage threshold and conversion efficiency for biomedical applications. *Nano-Micro Letters, 16*(1), 122.

23 Zhang, B., Wu, H., and Jiang, X. (2022, November). A laser-coupled dual excitation intravascular ultrasound transducer for sonothrombolysis. In *2022 IEEE Nanotechnology Materials and Devices Conference (NMDC)* (pp. 37–40). IEEE.

Index

Note: Page numbers in *italics* refer to figures, and those in **bold** refer to tables.

www.ingramcontent.com/pod-product-compliance
Lightning Source LLC
LaVergne TN
LVHW011519200926
848484LV00001B/73
* 9 7 8 1 3 9 4 3 6 5 0 9 8 *